Flying Birds

Flying Birds

DAVID AND KATIE URRY

foreword by Peter Conder

HARPER & ROW PUBLISHERS New York and Evanston

First published in England by Vernon & Yates Ltd.
under the same title
Layout by David and Katie Urry
FLYING BIRDS
Copyright © 1969 by David and Katie Urry
All rights reserved
Printed in Great Britain for Harper & Row, Publishers, Inc.
No part of this book may be used or reproduced
in any manner whatsoever
without written permission except in the case
of brief quotations embodied
in critical articles and reviews.
For information address
Harper & Row, Publishers,
Incorporated, 49 East 33rd Street,
New York, N.Y. 10016.
Published simultaneously in Canada by
Fitzhenry & Whiteside Limited, Toronto.

FIRST AMERICAN EDITION

Library of Congress Catalog Card Number: 74-110974

Foreword

I was very honoured and pleased when David and Katie Urry asked me to write the foreword to their book. I had first seen their work when David Urry won first prize in the 'Flying Free' Exhibition organised by Kodak Limited and the RSPB, and Katie won the second prize. At that time I was most impressed by their photographs which combine beauty with technical excellence.

The book before you combines superb photographs with a text that helps one to understand and appreciate the mechanics and beauty of bird flight, for the Urrys have taken as their theme 'The beauty of the bird in the air'. Dr Urry is a trained zoologist who has become a professional photographer and the text reveals his knowledge and understanding of the dynamics of bird flight.

There are two ways of looking at this book. You can read the text and study the pictures so that you can understand how the bird's body, wings, tail or feathers actually function as a bird flies, hovers, glides or alights. Or you can look at the photographs picking out the shape and lines created by the birds, the feather patterns or wing shapes, all of which create interest and beauty in a picture.

I felt after looking at this collection of photographs that a new world of visual exploration was being opened to me. I had previously felt that a certain staleness was coming to bird photography; that the field of pin-sharp nest photography has been explored to perfection by Eric Hosking and his contemporaries; that too many photographers were trying to imitate them rather than exploring new compositions, new bird subjects, different methods of processing to provide new print textures. On the other hand the Urrys are exploring pictorial effects of all sorts of things which were at one time considered to be technical faults, such as blurred outlines or grain. Yet I do not wish to give the impression that the majority of the photographs are blurred or grainy. They are not. In some photographs the Urrys have used the blurred wing-tips as an indication of movement, which is really how the human eye more often sees a bird's wing and which was how Lillieffors, the Swedish artist, often painted birds. In others, by freezing a bird's movement with high speed photography and by choosing a time of day when the bird was flying past them in line with the sun, they show the patterning formed by the feather webs. By choosing a low evening sun they have photographed the sculpturing of the underwing, creating a pleasurable pattern of blacks and greys and whites.

Another feature is the shape revealed by the postures in which the birds have been photographed. I have the feeling that the Urrys have consciously selected from their collection of photographs those which show interesting lines; in many it is worth letting your eye follow the line of the forewing round the head and down the other wing. Sometimes it is the shapes of the birds on the white background which pleases. For instance, see the twisting shags on page 94.

Again it may be repetitive effect seen, for instance, in the photograph of gannets on page 83. The wings are all almost in the same position. The picture reminded me of the paintings of Piet Mondrian, many of whose paintings developed from the lines of branches and twigs of trees and the spaces between them. You will find the same theme recurring in many other photographs on these pages. And finally grain, which is used particularly effectively for the introductory photograph in the section on 'Winter Shores and Marshes' and again for 'Winter in the City', here reminds me of the 'pointillisme' technique employed by the French painter Seurat.

I think that generally neither the bird artist nor the bird photographer has explored the shape and colour of birds to the full extent that their various media allow. If one examines either modern paintings or creative photographs in other fields one can see enormously variable types of exploration, some pleasing and some not. That this type of exploration has not taken place in the ornithological field in Britain may be due to the 'scientific' approach of the birdwatcher who, because of current emphasis on looking at detail in a bird's plumage, is unable to look at it through half-closed eyes as a beautiful object. This book will give birdwatchers an opportunity to take a new look at birds and see that they are beautiful.

PETER CONDER
Director of the Royal Society for the Protection of Birds

Contents

FOREWORD	7
FLYING BIRDS	11
HOW BIRDS FLY	12
SUMMER SHORES AND ISLANDS	22
Puffins	24
Razorbills	36
Guillemots	48
Arctic terns	58
Fulmars	74
Gannets	80
Shags	92
Kittiwakes	96
Lesser black-backed gulls and herring gulls	108
WINTER SHORES AND MARSHES	120
WINTER IN THE CITY	148
WINTER FIELDS AND TREES	168
PHOTO-TECHNIQUE	189
INDEX OF BRITISH, AMERICAN AND SCIENTIFIC NAMES	192

Flying Birds

Not all birds fly, nor are all flying animals birds, but the flying bird has a special place in the mind of man. To understand something of flight will not detract from the aesthetic enjoyment of birds; the wild appeal of a skein of geese punctuating the early sky or the stately flight of a gannet are enhanced by a realisation of the principles governing their progress through the air.

Once a bird takes to the wing it is exercising a function to which almost every aspect of its body is attuned. To watch a bird fly is to witness the result of a long and rigorous process of perfection that has taken place through countless generations of evolutionary change. Natural selection does not tolerate the inept or clumsy.

There are many different types of flying birds, each variety specialised to a particular kind of life in a certain environment; these different requirements have produced idiosyncrasies of flying behaviour and many variations in wing shape.

Natural laws have shaped and moulded the flying bird; structure has been dictated by function. The result has been the creation of animals that, when flying, find few parallels in beauty and graceful motion.

How Birds Fly

It will be obvious that this book does not set out to be an exhaustive treatise on bird flight: our concern is with the beauty of the bird in the air. Since this is a functional beauty, it would nevertheless be inappropriate to dispense with all technicalities, for herein lies the key to much of the aesthetic pleasure derived from the flight of birds: each shape, contour and movement has been refined until it has achieved a fitness for purpose and an economy of structure that results in the elegance our eyes behold. It is our intention to set out as briefly as possible some of the basic principles that may help towards an understanding of the exhilarating spectacle of bird flight.

To achieve flight an object has to overcome the gravitational forces urging its return to earth. One method of flight is to travel through the air, deriving a lifting force from the air that is passing. In an aeroplane the engines supply the propulsive force, the wings providing the lift. The bird's wing performs both duties and is correspondingly more complex.

A wing is, in fact, an aerofoil, a name given to any structure that, when moved through the air in a certain way derives a lifting force from the action of the air.

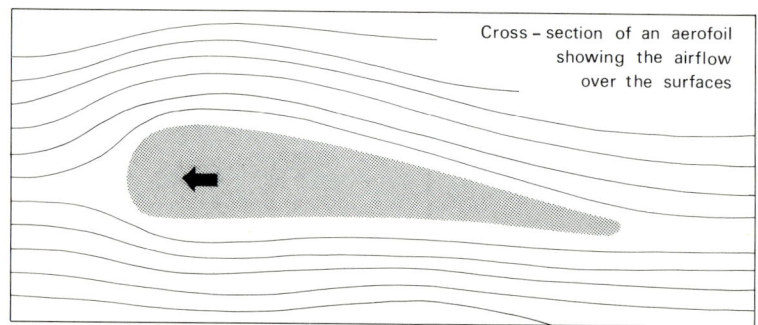

Cross-section of an aerofoil showing the airflow over the surfaces

When an aerofoil is moved through the air, air will pass over its surfaces. The air flowing over the upper surface of an aerofoil has further to travel, and, therefore, moves at a greater speed; this results in less air pressure on the upper than on the lower surface, the difference in pressures producing the lifting force. Various factors can influence the lifting ability of a wing; among these are the area and shape of the wing, the speed at which the wing is moving, and the angle of attack.

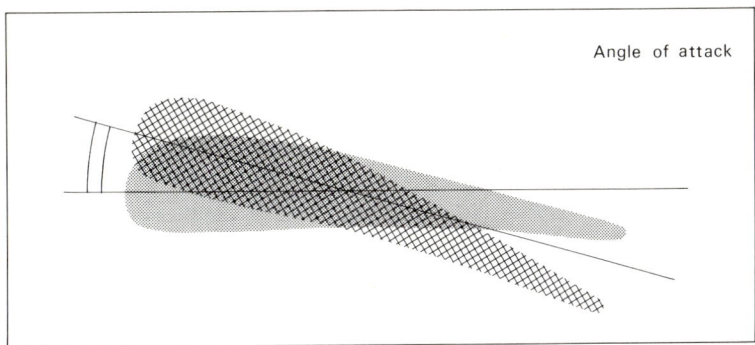
Angle of attack

If the angle of attack is increased, so are the lifting properties of the wing. However, this gain cannot be exploited indefinitely, for when a certain angle is reached the air no longer flows smoothly over the upper surface of the wing, but becomes turbulent, the wing loses its lifting properties, and the wing is said to stall.

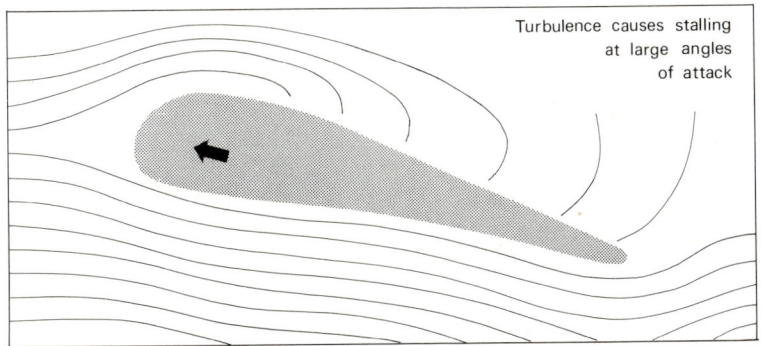
Turbulence causes stalling at large angles of attack

A bird's wing makes this condition of stalling easy to observe, for the turbulent air over the upper surface ruffles the feathers. There are various ways by which the tendency of a wing to stall can be reduced, one of which is to position a small subsidiary aerofoil in front of the main wing. Air is then actively directed over the upper surface of the wing, and a smooth airflow is more easily maintained. This device is known to aeronautical engineers as a Handley-Page slot, the counterpart of which in a bird's wing is a small movable structure, the bastard wing.

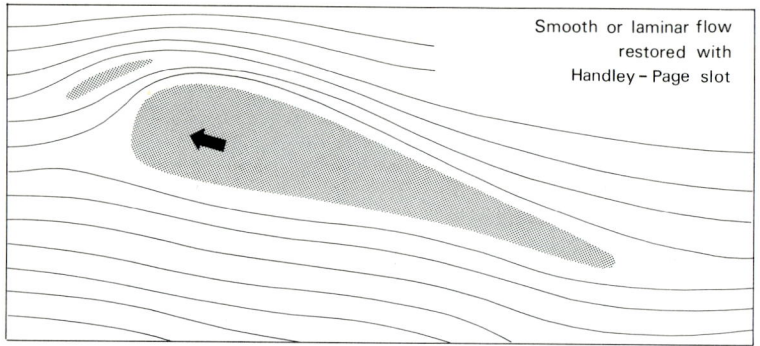

This principle can be repeated a number of times so that the wing becomes a series of narrow aerofoils, familiar as the divided or slotted wing tips of such slow-flying birds as eagles, crows and rooks. In many other birds the same effect can be achieved temporarily by asymmetrically shaped feathers, forming a series of slots when the feathers are fanned out.

Lift is not the only force that a wing experiences. The resistance of the air also gives rise to a force known as drag, which tends to slow down the forward motion of the wing. This drag must be overcome, either by propulsive force, or by off-setting it against the force of gravity by descending. Because there is no single wing design that is ideal for all purposes, birds of different flying habits have dissimilar wings. A large wing area gives a good lifting force, but also carries the disadvantage of causing a lot of drag; as a result, wings of large area are found on birds such as eagles, buzzards and short-eared owls, that fly slowly, quartering the ground for their prey. Another important factor is the wing span, for birds with a large wing span need relatively less energy to maintain them in flight. By possessing a long, narrow wing a bird may have a relatively large wing span without a proportionate increase of the area of the wing, or the drag. The albatross shows one of the best examples of a long, narrow wing in a fast-flying bird.

Wings that are long and narrow, like those of the albatross, or the needle-like wings of a swift, are said to have a high aspect ratio (the ratio of the wing span to the breadth of the wing). Conversely, a short, broad wing has a low aspect ratio. Long, narrow (high aspect ratio) wings suffer from the disadvantage that they stall at relatively fast speeds, and so would be of no use to a bird like a vulture, slowly soaring over the countryside; consequently birds with these flying habits have evolved low aspect ratio wings. Low aspect ratio wings can be seen in the photographs of rooks.

If the photographs of rooks are compared with those of gannets it can be seen that not only do they differ in aspect ratio, but also in shape, the gannet having a pointed wing tip, the rook possessing a blunt-ended wing with well separated primary feathers. This is again related to the speed of flight, for a pointed wing induces minimum drag but stalls first at its tip, and would be unsuitable for slow-flying birds.

Matters are not quite as simple as this in a bird's wing. If the photograph of a gannet flying at speed is compared with the photograph of a landing gannet, the wings present a very different appearance; by means of feather movement the gannet's wing is approaching the form of the rook's wing as it flies in slowly to land. The aspect ratio has been lowered, there are even wing slots at the wing tips and the bastard wing has been brought into action to reduce further the stalling tendency of the wing.

So far, in the interests of simplicity, the bird's wing has been considered as a fixed wing, like an aircraft wing. This, of course, is not the case, for the bird's wing also has to provide the power for flight by flapping.

Wing flapping consumes a considerable amount of energy, and many birds do, at times, conserve this energy by gaining the power for flight by another means, soaring. Unpowered flight over considerable distances is possible, a fact made familiar by the well-known achievements of long-distance glider pilots.

The glider pilot manages to fly only by balancing the gravitational pull of the earth against the drag forces acting on his plane; the only way he can do this is to incline his glide path downward. In perfectly still air he would not get very far, but fortunately the air is seldom still over land. This is due to the relative warming or cooling of air over such features as mountains, forests, rocks or buildings, the warm air becoming less dense, and the cool air becoming more dense. Warm, light air rises as thermal currents, and if these are strong enough, although the glider is flying downward relative to the air, this will be offset by the thermal and it will be rising relative to the land.

Birds that soar over land, such as eagles and vultures, employ the same tactics, rising on upward thermal currents. This type of behaviour imposes certain features on the wings of such birds. They have a large wing area, giving maximum lift, so that

only a shallow diving angle is necessary to compensate for drag, and the slow rate of sinking is easily offset by a rising thermal. It is also necessary for them to be able to manœuvre easily and so remain within the often small confines of the thermal; this is facilitated by short wings. There is, then, good reason for the relatively short, broad wings of these birds.

Only under very exceptional circumstances are thermals set up over the sea, since the temperature of the sea is uniform over very large areas. Nevertheless, marine birds are not without means of soaring, for it is a feature of the oceans that strong, constant winds are often blowing. Upward currents of air occur where wind is deflected by waves, and these upcurrents are utilised by such birds as shearwaters; they gain height soaring in the upcurrent from one wave, and then glide down to meet the upcurrent from the next wave.

Some sea birds show another type of soaring behaviour. When wind blows over the sea, the lower layers of air are slowed down by friction between sea and air. Anything rising away from the sea surface will, therefore, climb into steadily stronger wind, until it reaches a height of about 100 feet where this effect usually dies out. A bird travelling well above its stalling speed is not obliged to glide downward; it can climb, but with loss of airspeed, and only until its stalling speed is reached. If it is over the sea it will climb into stronger winds, and if it is headed into wind this will compensate for the loss of airspeed; the bird will, however, lose ground speed, and so will not move far during this operation, which is essentially one of gaining height. At about 100 feet this advantage can no longer be gained, and at this point the bird turns downwind, entering a shallow dive, and thus gaining both air and ground speed and covering a considerable distance over the sea. By the time it is almost down to sea-level it is flying fast, and then turns into the wind, repeating the cycle of events. The albatross is the acknowledged master of these tactics, and since this type of soaring demands a high gliding speed the birds have high aspect ratio wings.

Another form of soaring can be seen where the junction between land and sea is marked by cliffs. Wind is deflected upwards by cliffs, and birds from both land and sea take advantage of this rising air. Buzzards, eagles and gulls can be seen soaring over the same cliffs, but the greatest virtuosity in cliff soaring is displayed by the fulmar.

Only certain birds, under these special circumstances, can derive the power for flight from soaring. Soaring birds cannot soar all the time, and most birds cannot soar at all; the power for flight must then be provided by the bird itself, by flapping the wings. Once the wings are flapped they are performing a dual function, behaving both as aerofoils and propellers. The form of the wing during its beat is constantly changing, due to both muscular movements of the bird and the bending of the feathers by the air. The factors that produce the lift and thrust are complex, and not yet fully understood.

There are various different types of flapping flight, and only a brief description of major types will be attempted here. When pigeons, owls, ducks and many sea birds are either taking off or landing, they perform a complex type of flapping.

The wings are not simply flapped up and down, but halfway through the downstroke they are moved forward, almost meeting in front of the bird. They are then drawn rapidly backwards, followed by extension above the body ready to begin another downstroke. As the wings are brought down they press on the air and cause the bird to be lifted; during their forward movement they act as aerofoils and provide more lift. A powerful forward thrust is developed from drawing the wings backwards and upwards. Lift is developed mainly on the downstroke, and the bird is propelled forward largely by the upstroke of the wings. This method of flight, used by some birds during take-off and landing, cannot be sustained for long periods. A bird has two main sets of flight muscles, a large set supplying the power for the downstroke, and a very much smaller set responsible for the upstroke of the wing. In this type of flight the upstroke, giving forward thrust, uses a lot of muscular energy, and the small muscles responsible for this would soon become exhausted.

When take-off is achieved, and a fair flying speed attained, the whole pattern of wing movement alters. The wings are now flapped up and down in a much more simple manner. Lift is given all the time by the aerofoil properties of the wings, but only the downstroke is powered by muscular effort, producing forward propulsion as well as lift; the upstroke is one of passive recovery, hardly using the small flight muscles. This type of flight can be sustained for long periods of time, nearly all the muscular effort coming from the large, strong, principal set of flight muscles.

Most large birds, then, have a rather complicated take-off and landing flight, and a simpler and much less exhausting method of level flight. Most small birds, such as sparrows, finches and tits only use the simpler type of flight and do not have a basically different type of wing movement during take-off and landing; they simply move the wings faster through a greater arc when landing and climbing. Some large birds, such as the magpie, fly in a similar manner to the small birds, and do not have two patterns of wing movement. Some very large birds, such as albatrosses and swans, have no special take-off flight, and have to flap along until they reach the speed at which they can fly. It is easy to see that the long, laboured flapping of a swan taking off is very different in nature from the almost vertical, springing of a duck.

To summarise, all birds in level flight use muscular effort only for the downstroke, but most large birds have a different pattern of flight during take-off and landing, entailing the use, for a short time, of the smaller set of flight muscles to power the upstroke.

It must also be remembered that the tail is an important structure in flight. It can be used to produce a tilting of the body, and also as an air brake when landing. Many of the photographs show landing birds with the tail feathers spread out to their fullest extent for this purpose.

Flight is a very demanding method of locomotion, requiring a great deal of power. This power is provided by the flight muscles of the bird, which are able to produce more power, weight for weight, than the muscles of many other animals.

Nevertheless, it is important that this energy should be used efficiently. Energy will be wasted if the bird is unduly heavy, for this unnecessary weight must be lifted off the ground and maintained in the air. The heaviest part of the body is the skeleton, but in birds this weight is reduced to a minimum. Some bones have fused together to give maximum strength, allowing a measure of reduction in the amount of bone present. Many bones are hollow and contain air sacs, some braced internally, often in a way similar to the trusses strengthening aeroplane wings.

Not only is it important for a bird to be as light as possible, but flight will be more controllable if the existing weight is concentrated as near the centre of the body as possible. The compact streamlined shape of a bird achieves a smooth air flow over the body. Teeth are heavy things situated well away from the centre of gravity. Birds have done away with teeth, but still need to grind their food, a function now carried out by stones and grit in the gizzard, less of a disadvantage being near the centre of gravity.

In spite of all these economies in the needless expenditure of energy, great demands are made on the metabolic activities of the bird, and these therefore possess a corresponding degree of efficiency. Chemical activity takes place faster at high temperatures, and birds have a body heat of 40 °C (104 °F), allowing the biochemical reactions within the tissues to proceed rapidly. The bird's lung is of a very efficient type. Our own lungs contain a lot of residual air that is not breathed in and out, so that they are never completely ventilated. When a bird breathes, air passes through the lungs into air sacs, and there is a complete change of air in the lungs with each breath. When a bird is flying the flight muscles squeeze the rib cage, effectively pumping the air in and out of the body.

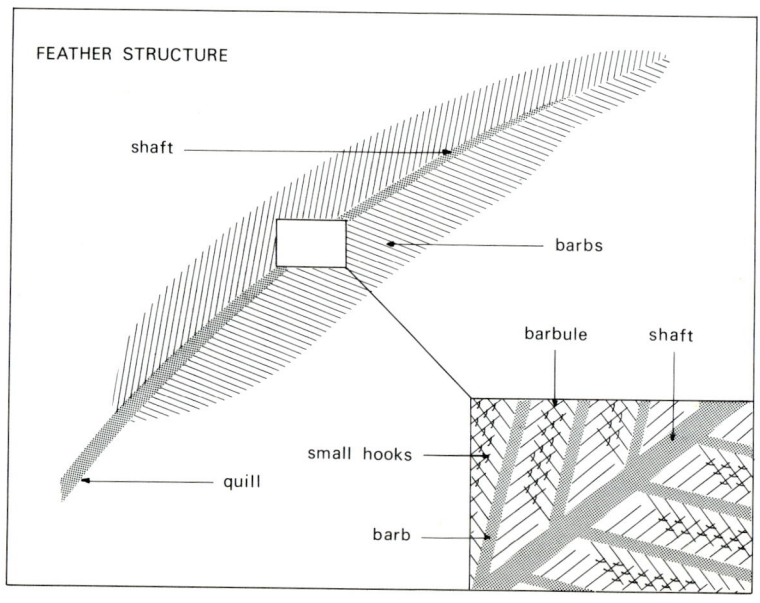

HOW BIRDS FLY

One of the most important features aiding a bird to fly is the possession of feathers. These are proverbially light, yet are amazingly strong structures. They provide the major part of the aerofoil surfaces of the wing, and also mould the body contours to the required streamlined shape; they also form an exceptionally good insulation, important to an animal with a high body heat that often has cold air rushing over it.

The barbules of the feathers have small hooks on them, which interlock with those on adjacent barbules. If the feathers are ruffled, this device enables the bird to smooth them back easily to their original arrangement by preening. The down feathers of chicks do not have these hooks, accounting for their fluffy appearance.

The main feathers of the wing and the feathers of the tail are known as the flight feathers. The primaries are on that part of the wing corresponding to a hand, the secondaries on that part corresponding to a forearm. Overlying their bases are the upper and lower wing coverts, feathers which give the wing the desired aerofoil shape.

Summer Shores and Islands

Flying birds can be seen with no greater ease or enjoyment than at colonies of sea birds in spring and early summer. Some of these sea birds maintain tolerable relations with land at other seasons, but others are shy of land and their instinctive desire to remain over the vastness of the ocean is only overcome by the stronger demands of reproducing their kind.

For an all too brief period the cliffs and shores are enlivened by this throng. Cliffs that have hung lonely over winter seas become filled, ledge upon ledge, with the ceaseless activities of these birds. Below, the sea is carpeted with resting birds, to the horizon the sky is interlaced with the traffic of wings, and the air is filled with the harsh, wild sounds of summer shores and islands.

Puffins

SUMMER SHORES AND ISLANDS

It is in March or early April that the puffins come to land; their winter has been spent at sea, but now they come to nest in their colonies of burrows, and until they leave in July or August their friendly, often comical, presence will enliven the shores and islands. Parties of puffins will gather on rocks or grassy slopes, they will be dotted on the shimmering summer sea, or flying fast and direct about their business with the slightly agitated expressions of commuters.

The wings are small in proportion to the plump bodies they have to carry, and in consequence the puffin has to fly fast, and with quickly moving wings, the wing beating approximately 12 times per second. The small wing area is, however, a great advantage to the puffin, since the wings are used for underwater swimming when fishing.

When coming in to land the small braking area of the wings is supplemented by the spread-eagled orange feet and the fanned tail feathers. In calm weather the puffin lands quickly and easily, the body suspended beneath whirring wings, with the neck and head craned forward. On windy days they have to struggle in, constantly compensating for variations as the wind is deflected by the rocks. All attempt at dignity is abandoned when landing on the water, the birds splashing down awkwardly and often completely submerging.

To take off from the water the puffin has first to gain its flight speed, and patters across the water with wings and feet for some distance before rising. From a cliff-edge the flying speed is attained by gliding down on stiffly held, rapidly vibrating wings, a type of flight which is sometimes used as a display.

In spite of their small wings the puffins are competent fliers, travelling at 50 miles per hour or so, and ranging for up to 50 miles from the colony in search of fish.

One of the most memorable spectacles the puffins provide is at dusk on calm evenings, when there is enormous activity. Whole bays are filled with puffins flying and wheeling at high speed, endless streams of small black shapes forming a silhouetted pattern of counter-movement against the delicate colours of a summer evening.

A puffin coming in to land. The wings are being brought forward in a braking action, the primaries bent back by the resistance of the air. The braking area of the wing has been increased as much as possible by fanning out the flight feathers. This alteration in the wing shape will also derive maximum lift and give a low stalling speed; the tendency to stall is further reduced by wing slots at the wing tips and between the raised bastard wing and the main wing. This bird is very near to stalling; the air flow over the upper wing surface has become turbulent, and the upper wing coverts are ruffled on the bird's right wing. Maximum lift at low speed is also being gained by holding the wing at a steep angle of attack. The body is no longer horizontal, but is now hanging at an angle from the wings, with the feet ready to touch down.

SUMMER SHORES AND ISLANDS

Puffins often assemble in numbers on rocks adjacent to their colonies. Periodically they suffer disturbance from a passing gull swooping low, when they all take off and fly in a circuit before returning. Various stages of the movement of the wings through a large arc during take off can be seen in this photograph.

A combined braking action and sharp turn. Notice how the puffin's primaries are bent by the pressure of the air. Further braking and control are being obtained from the spread tail feathers and extended feet.

SUMMER SHORES AND ISLANDS

Two views of puffins coming in to land. A slight adjustment for direction is being made by the bird in the lower picture, and consequently the wings appear asymmetrical. There is a large angle of attack in both cases, and the bird in the upper picture is aiding braking with its tail.

This is the normal flight attitude of the body in relation to the wings during level flight. This puffin is climbing slightly, having just taken off, and until more flight speed is gained it still has the slots at the wing tips and the bastard wing slightly raised to reduce stalling tendencies. The feet are now held back to reduce drag.

A puffin turning, but at fair speed. The feet and tail are being used as rudders, but the wing shape shows that the bird is neither braking nor travelling very slowly. The wing is not fully fanned out, the bastard wing is not raised, and there is a small angle of attack.

SUMMER SHORES AND ISLANDS

An early stage in landing, with the puffin looking for a suitable site. The feet have been lowered and the body is beginning to swing down from the horizontal. There is a large angle of attack, giving a tendency to stall, as can be seen from the ruffled upper wing feathers. The bastard wings are in full use to reduce stalling.

Two views of the normal, unhurried flight used by puffins when launching from the cliff-edge. The wings are held out stiffly, and rapidly vibrated through a very small arc, as the bird gains speed by inclining the flight path downwards. A curious feature of this flight is that the feet are held in a crossed position.

SUMMER SHORES AND ISLANDS

Another group of puffins that have just been disturbed from a rock assembly. These have gained some speed, and the flight is changing to one in which the wings are moved through a smaller arc.

SUMMER SHORES AND ISLANDS

Puffins adopt a variety of dramatic, and often amusing, postures as they come in to land.

Razorbills

SUMMER SHORES AND ISLANDS

Intermingled with the puffins on the sea and in the air around the large sea bird colonies is another auk, the razorbill. However, once on land there is segregation, for although both may be seen together sunning themselves on a rock, nesting is carried out in different sites; whereas puffins nest in colonies of burrows, razorbills nest on the cliffs, on ledges, in crevices or under boulders. The term 'nesting' must not be taken too seriously when referring to razorbills, for they do nothing in the way of constructing a nest.

Like the puffin, the razorbill has a small wing area for its size, for again the wing shape is a compromise between the requirements of flight and swimming under water. This results in a fast, direct flight with quickly beating wings, typical of auks. Nevertheless, there are differences in detail between the flight behaviour of puffins and razorbills.

Very often razorbill parties can be seen flying in a rough line-ahead formation, adding yet another pattern to the activity in the sky.

Occasionally razorbills, often in pairs, can be seen flying in a way quite different from their normal flight; now, instead of beating quickly through a shallow arc, the wings beat quite slowly, and are moved through a large arc. One can only presume that this is some form of display.

This type of slow wing action can be seen more regularly when the razorbills launch themselves into the air from a cliff. They gain the necessary speed for flight by flying downwards, moving the wings slowly and stiffly in much the same way. The wing beat then changes to the normal quick beating motion as the bird levels out and flies off low over the water.

Whereas a puffin usually drops into its landing place, a razorbill normally sweeps in an upward arc to its ledge. This climb will, in itself, cause a loss of speed, and the final braking is achieved by the wings being flapped forwards in a scooping motion, assisted by the outstretched, spread feet and feathers of the tail.

Razorbills come in to land rather faster than puffins, and consequently only approach a speed near stalling during the last moments of the landing flight. This bird is braking hard prior to landing; the body has swung to an almost vertical position so that the powerful downstroke of the wings is now forwardly directed, giving maximum braking. Further braking is being derived from the partially spread tail feathers and the extended feet.

SUMMER SHORES AND ISLANDS

A side view of the razorbill's landing procedure. The forward stroke of the wing has just commenced, the wing tips bent back by the resistance of the air.

The normal, level flight of the razorbill. The wing shape of a bird is not a fixed entity, and here the razorbill has a narrow, pointed wing, suitable for fast flight.

A razorbill approaches a landing-site, turning as it sweeps in.

SUMMER SHORES AND ISLANDS

A razorbill flying past and turning as a landing-site is prospected. The feet are lowered here mainly as steering aids in the fluctuating air currents over the cliff. There is only a low angle of attack, as the bird is travelling quite fast and deriving sufficient lift.

An early stage in landing. The body of the razorbill has swung to the near vertical and the wings are about to be brought forward; the feet are out, but the tail has not yet been fully brought into action as a brake. The slotting of the primaries is well shown.

A fairly late stage in landing. The razorbill's speed has been reduced, stalling tendencies minimised by wing-tip slots and the raised bastard wing, and the aspect ratio has been altered by fanning out the flight feathers, so broadening the wing. Tail and foot brakes are firmly applied.

SUMMER SHORES AND ISLANDS

A razorbill landing, with the tail not yet brought into use as a brake.

A razorbill overhead, turning very slightly. The abrasion of the tips of the primaries, usual at the end of the breeding season, is very noticeable in this picture.

SUMMER SHORES AND ISLANDS

This photograph of a razorbill gives some idea of the exertions of landing, and also shows well the large area of resistance that can be made by wings, body, tail and feet.

Turning after a preliminary 'dummy run' over a landing-site. The razorbill's feet and tail, although spread, are not so flexed down as in a landing bird; they are more in line with the body and are acting as rudders.

SUMMER SHORES AND ISLANDS

A razorbill inspecting the landing-site. Very often the birds will make one or two circuits before actually landing. The spread tail and feet are aiding steering, and also giving some extra lift at the tail end. At a fair speed such a low angle of attack causes no stalling and the upper wing is quite unruffled.

A razorbill approaches a landing-site after a successful fishing trip.

Guillemots

SUMMER SHORES AND ISLANDS

The third of the commonest British auks is the guillemot. Guillemot colonies are generally found in the same localities as those of razorbills and puffins, but there is again a difference in nest site. Like the razorbill there is no nest, but the guillemot lays its eggs on the most exposed and inaccessible ledges of the cliffs, and on ledges or platforms at the cliff-top, or on top of rock stacks.

The type of flight and the flight behaviour are much the same as those of the razorbill, but the flight of the guillemot, due to its more perfectly streamlined outline, gives the appearance of possessing an elegance denied to razorbills or puffins.

The method of landing is largely decided upon by the situation of the nest. Those nesting on ledges will sweep up to them in the same way as a razorbill. Landing is a problem to those nesting on the top of stacks, as these are intensely crowded, and the guillemot has to drop down into a space only a little larger than its own body. The action used is very reminiscent of that of a puffin, but false runs are not uncommon, when the bird has to continue to another circuit before repeating the attempt.

A guillemot turning. The feet and tail are extended to act as rudders, although in this particular bird the tail cannot be very effective. The wing is just beginning its downstroke, and is fully extended so that the maximum power is derived from this movement.

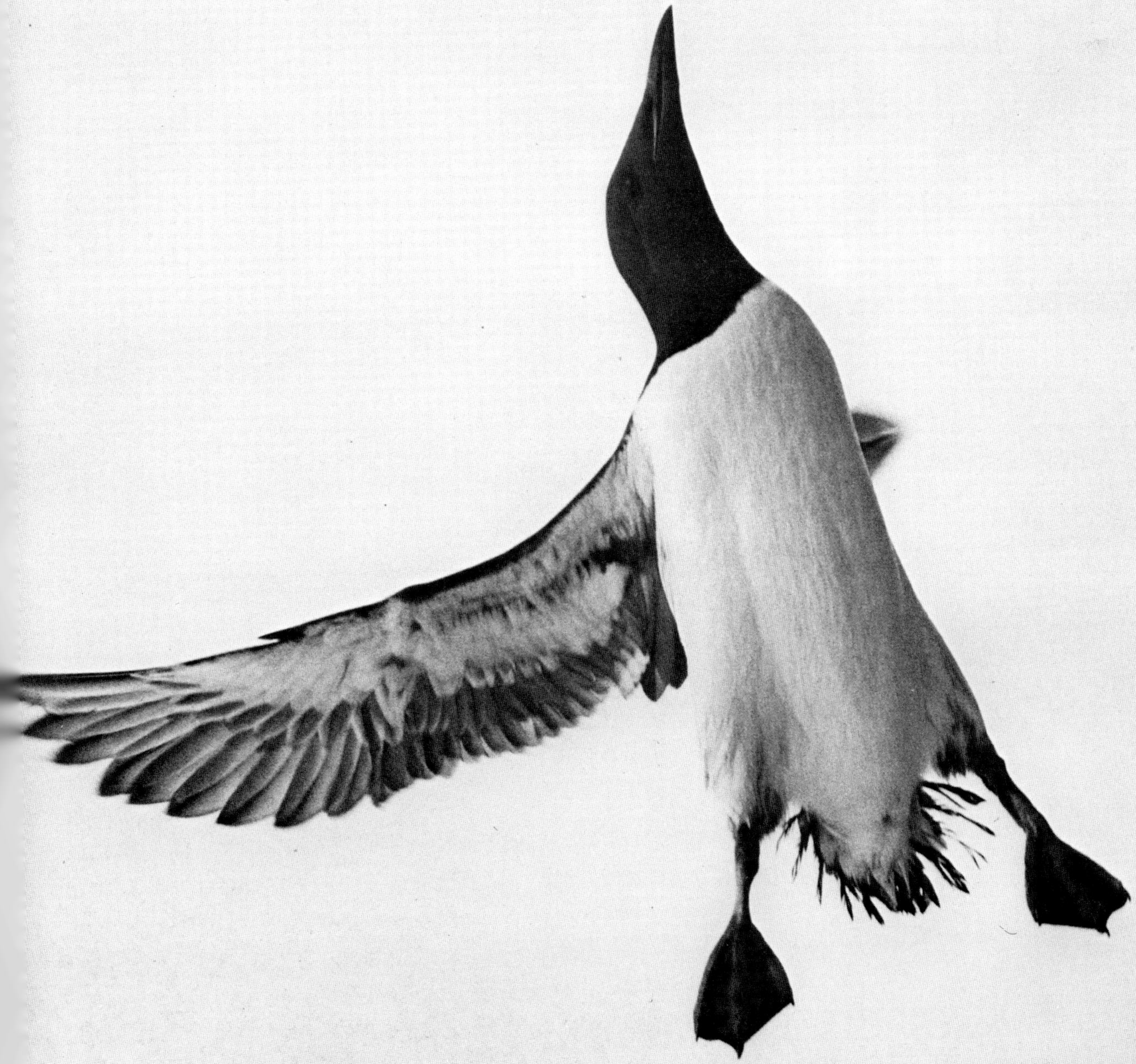

The wings of this guillemot are seen during the upstroke, when they are partially flexed. The upstroke of the wing in level flight is a passive recovery movement, and is brought about by the action of the air, not by muscular effort.

SUMMER SHORES AND ISLANDS

Here the guillemot's wings are nearing the top of the upstroke and are being extended for the powerful, muscular downstroke.

The wings are right at the top of their arc, just about to begin the downstroke. The guillemot has over-run its landing on a rock stack and is gaining speed to make another circuit. It was photographed as it flew over the main cliff, where in the variable air currents extra control from the feet and tail are still needed. Once it has gained full speed over the sea the tail feathers will be folded to a streamlined point, and the feet will be tucked in line with the body. The head, body, tail and feet of the guillemot then form a very well streamlined shape, an important asset to a fast-flying bird.

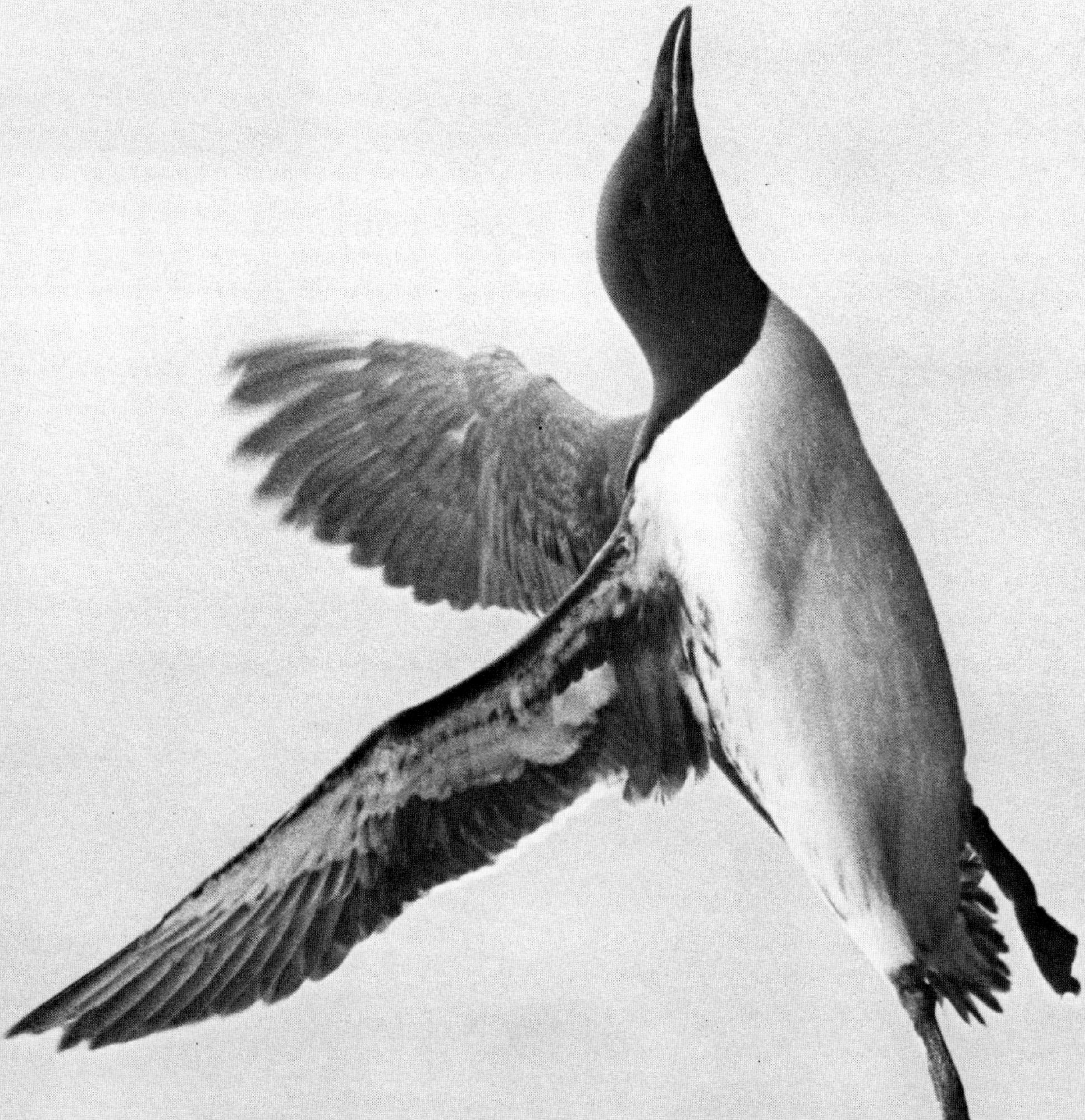

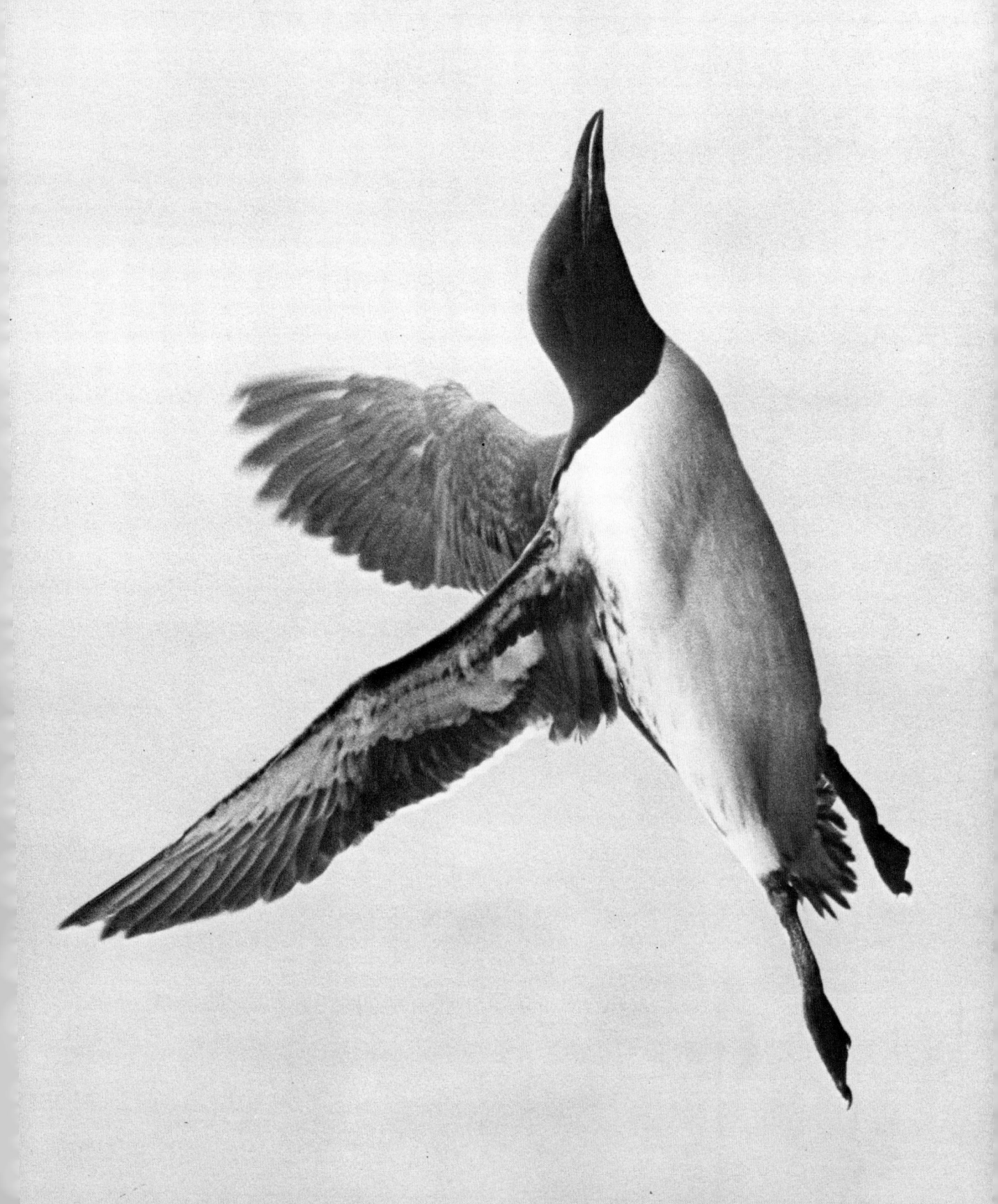

These photographs further illustrate the difference between the upstroke and downstroke of the wings. Three of the guillemots are in various stages of the downstroke, with wings fully extended, whilst the fourth, with partially flexed wings, is seen during the upstroke. The varied positions of tail and feet show how these structures are used for control in conjunction with the wings.

SUMMER SHORES AND ISLANDS

Arctic Terns

SUMMER SHORES AND ISLANDS

The terns provide a complete contrast to the auks. Terns have a delicate ease and grace in their flight that is virtually unrivalled among sea birds. In normal flight the wings beat about 3 times per second, in a slow, purposeful manner, and the effect of lightness and buoyancy is increased by the bird rising as the wings are brought down, and sinking slightly on the upstroke. Not only are the terns birds of distinctive beauty in the air, they are also some of the most versatile, having various types of flight which play a considerable part in their social behaviour.

When fishing the tern flies slowly, looking down into the water, occasionally glancing forward to check that the course is clear. On sighting a fish the tern dives with partly closed, angled wings, plunging like a white arrow into the sea. Sometimes, if in doubt, it will not dive immediately, but will hover over the spot where it suspects there is a meal; having made up its mind it will then either dive, or move slowly on in further search.

The hovering flight is best seen over the colony. A tern seldom drops straight to its nest, but, often after losing height by gliding on steeply angled wings, hovers for a short while before descending. The hovering tern is a fascinating sight, the partly translucent wings are now moved much faster, 7 or 8 beats per second, and the form of wing beat is quite different from that of the tern in level flight. Now the wing traces an arc similar to that employed during landing and take-off, the head nodding up and down as it checks its position, and the tail constantly combating any attempts the wind may make to upset the bird's stability. Sometimes the tern appears to stop its wing beat momentarily, to fall for a few feet, and to hover again lower over the nest. In positioning itself over the nest the tern can move horizontally both forwards and backwards; when reversing the wings thrust forward right over the head during the downstroke. If another tern flies too near, the hovering bird will immediately interrupt its attempts to land and give chase. In a high wind the landing behaviour is different, the birds balancing on the wind with angled wings, embarrassed by their buoyancy, as they slowly descend to the nest. Once the tern has alighted the wings are usually held up over the back for a moment, this lovely position being held longer in display if alighting near its mate.

Arctic terns have various patterns of display which involve flight. One of the most common of these involves two birds; they leave the ground with one bird leading, and when they have gained a few feet in height they rise together on wings held high and fluttering. At a very variable height, sometimes considerable, they break into a type of flight which is quite unlike their normal flying; they now move at great speed with the wings held stiffly and not quite fully extended, beating only through a shallow arc; each bird follows a curving zigzag path, resulting in the pair constantly changing sides, and the impression is reminiscent of two expert ice-skaters. Sometimes a near-by bird cannot resist joining in this spectacular display for a while. The descent from these excursions is usually by interspersed gliding and wing flapping, but occasionally the descent is a side-slipping downward rush. The final descent is an exaggeration of normal landing, with the wings held in a V above the back, giving the impression of a shuttlecock dropping to the ground.

Another type of display flight of the arctic tern involves the ritual use of food,

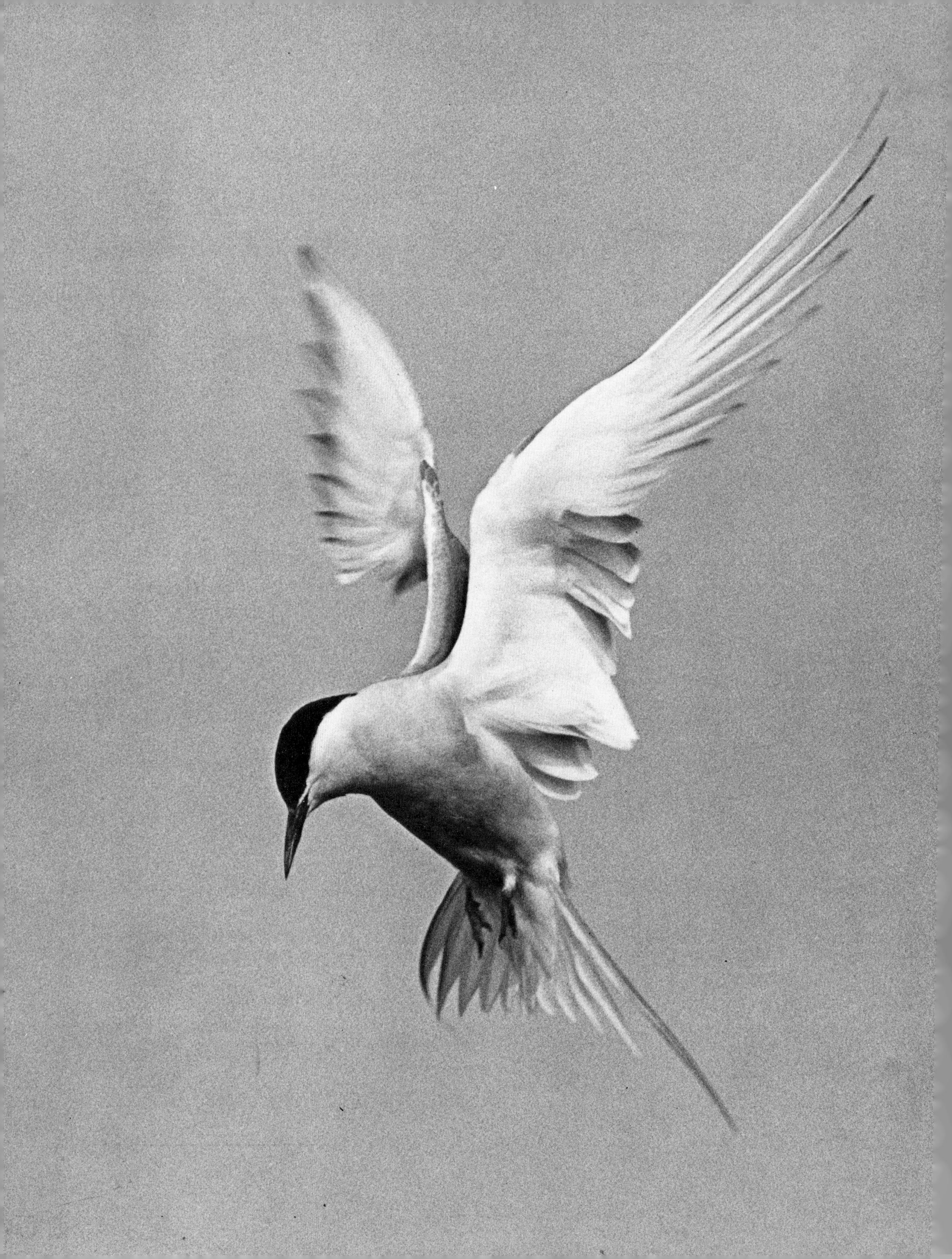

in this case fish. This display often begins on the ground when one bird, with a fish in its beak, adopts a display posture, with its beak pointed upwards, and its folded wings drooping so that they touch the ground. Both birds adopt this position for a while, and then the bird without the fish adopts a submissive pose with the head forward. The fish is then presented to this bird, who takes it and flies off, followed by the donor. The bird that now has the fish flies slowly, with frequent short glides, during which the wings are held in a V above the back; the bird that is not carrying the fish follows the fish-carrier, flying from side to side. This 'fish-flight' does not always begin on the ground, for sometimes the fish-carrier is joined by the second bird while on the wing, and occasionally two birds may follow the fish-carrier in this display. In fact, all these types of display flight in the arctic tern have many variations, so adding to their interest.

Nearly always over a colony there are pairs of arctic terns rising together vertically on fluttering wings in the numerous quarrels that are continually arising. The arctic tern is an aggressive and jealous guardian of its nest; as well as attacking its own kind it will fearlessly attack intruders (including humans); the birds hover above them, making repeated diving attacks, striking with the bill, and at times aiming droppings with great skill.

The tern colony is well able to deal with intruders by these means, and another type of behaviour involving flight, the 'dread', appears to have little to do with defence. When a 'dread' is about to occur, the already clamorous noise of the colony increases; most of the birds in the colony, or in a section of it, then take off apparently simultaneously, in complete silence. This is an unforgettable moment, when the raucous din the ears have become accustomed to is suddenly replaced by the whispering of rushing wings as the birds stream low over the colony and out to sea, wheeling in complete unison. Sometimes the pattern of the flock disintegrates and the return to the colony is immediate, normal life and vocal activity being resumed. At times, however, these 'dread' flights last for several minutes, the flock pattern breaking and re-forming in a high-speed series of swooping movements. During the 'dread' the wing beat is somewhat faster than in normal flight; there are approximately 5 instead of the usual 3 wing beats per second.

These 'dread' flights pose some problems to students of bird behaviour, for it is difficult to determine the stimulus that initiates them, and it is equally hard to find an explanation for their significance. One suggestion has been that the 'dread' represents a temporary reassertion of the flocking or migratory instinct at a time when reproductive behaviour is normally dominant. This may be so, but when the reproductive behaviour ceases and the terns leave the colony at the end of the breeding season, they do not depart as a large flock in this manner, nor does it explain the curious silence. We have wondered if the 'dreads' had their origin as a form of distraction display, but it is difficult to imagine a need for this in the behaviour patterns of such an aggressive species, capable of attacking intruders.

The immediate cause initiating a 'dread' is equally difficult to determine; if one has not occurred for some time the sudden appearance of a human form, or a boat, will serve to start a 'dread', but if the birds have recently returned from

Hovering terns have a complex wing movement that allows them to remain almost stationary in the air for quite a time, unlike the momentary hovering that gulls are capable of performing. The tail is depressed and fanned out for extra lift and control, the long outer feathers, or 'streamers', adding further elegance.

SUMMER SHORES AND ISLANDS

one, these factors will not induce the behaviour. The converse is equally true, 'dreads' commencing when it is impossible to detect anything that could have alarmed the birds.

Whatever the significance of the 'dread' it seems deeply rooted in the behaviour of arctic terns. If there is a strong wind blowing, of force 5 or 6, the tern colony reduces flying activity to a minimum. The display flights, such as the 'fish-flight', are not seen, and even the aggressive fighting flights do not occur; the few birds in the air are those leaving or arriving at their nests. However, 'dreads' still occur, although with far less frequency than is normal, and they seldom sweep out over the sea, the pattern breaking almost immediately and the birds returning to the ground.

A typical position of the arctic tern when hovering; the wings are on the upstroke, or backstroke as it has become, with the flight feathers bent sharply forward by the resistance of the air. The upper wing feathers are somewhat ruffled by turbulent air.

SUMMER SHORES AND ISLANDS

Just commencing the downstroke. The wing slots are formed by a twisting of the primaries due to the resistance of the air, as well as the fanning of these feathers by the bird.

An arctic tern directly overhead in normal flight. The feathers are now held so that the wing is slender and pointed at the tip, a shape suitable for fast flight. The aspect ratio of the wing in this position is much higher than that of hovering terns. The tail is not spread, the feathers overlapping so that two tapering forms are made alongside the outer 'streamers', a shape giving little drag. The feet are tucked away to reduce drag further.

SUMMER SHORES AND ISLANDS

A typical position as a tern hovers and looks down, either to check its landing-place near the nest, or into the water in search of fish. The translucence of the wings is a very attractive feature of the arctic tern.

There is always activity in the air over a tern colony during the breeding season, with birds constantly arriving and leaving.

SUMMER SHORES AND ISLANDS

The terns flying over the colony form perpetually changing patterns and shapes.

An arctic tern just about to commence hovering. The feet are aiding braking, but the tail, not yet spread, is still in the position in which it is held in normal flight.

SUMMER SHORES AND ISLANDS

A frontal view of an arctic tern hovering. The number of photographs of terns in which the bird is calling gives some idea of the amount of vocal activity that accompanies hovering.

Arctic terns streaming out over the sea in 'dread' flights.

SUMMER SHORES AND ISLANDS

Fulmars

SUMMER SHORES AND ISLANDS

Nothing is more restful than to watch fulmars wheeling to and fro against the face of a cliff on a summer's day. The wings are held stiffly out from the body as the bird soars on the upcurrents of air caused where the wind strikes the cliff. Occasionally the wings are slowly flapped, but for long periods the fulmar glides without any wing flapping. The fulmar is such a master of this technique that it makes the task appear deceptively easy. However, although the wings are seldom flapped, the detailed form and attitude of the wings has to be constantly altered to adjust for the variable, gusty conditions that prevail in the air near cliffs. The gliding fulmar has to maintain its flight speed by descending through the air, and can only avoid loss of altitude by gaining height in compensating upcurrents of air.

Sometimes fulmars can be seen hanging almost motionless near the top of a cliff, balancing on the updraught. When coming in to the nest they often appear to experience difficulty in touching down. Whereas auks have to undergo energetic braking movements, the problem for the fulmar is to lose sufficient lift to be able to land, particularly when headed into a strong wind.

Away from the cliffs the fulmar flies with a steady, measured wing beat, interspersed with periods of gliding, but over the open ocean it can again soar, using the currents of air deflected upward by the waves.

A fulmar turns away from the cliff-face. The bird is soaring, and encountering a strong updraught it has reduced lift by partially flexing the wings. The position of the primaries forms efficient wing slots, so reducing the tendency to stall.

SUMMER SHORES AND ISLANDS

The soaring skill of the fulmar is achieved by having a large wing area; broad wings would not suit its fast flight, and so the wing area is gained by a large wing span. The use of the tail in steering can be clearly seen.

A fulmar banking at speed while soaring on the wind. The tail is held in such a way that it can still function as a controlling aerofoil without causing undue drag.

Head-on view of a fulmar over the top of a cliff-edge. When balancing on, and exploiting, upcurrents of air the fulmar uses its tail and feet as very important parts of its aerodynamic apparatus.

Gannets

SUMMER SHORES AND ISLANDS

Gannets nest on islands, their colonies often containing enormous numbers of nests. The arrival at such a gannetry is exhilarating; the ground is covered by nesting gannets, spaced in an orderly fashion, and the air is filled with their guttural voices and the foul stench of guano and decaying fish. Above the colony hundreds of these large, dazzlingly white birds continually circle.

The gannet is a strong flier, and may fly up to 100 miles from the gannetry fishing. The fishing area of the colony has to be large to satisfy the appetites of thousands of these large birds and their offspring.

When travelling to a fishing-ground the gannets are usually only 30 or 40 feet above the water. If a strong wind is blowing, parties of gannets, either strung out line-ahead or in an unequal V formation, can often be seen flying fairly near to the shore. Under these conditions they are very close to the water, tilting from side to side on the wind deflected upwards by the swell, so that they are seen sideways at one moment, from above the next, resembling a line of pure white cardboard cut-outs drawn over the undulating leaden sea, flecked with the dull white of breaking wave crests.

When actually fishing the height of the gannet may be anything from a few feet to over 100 feet. The gannet engaged in fishing faces into the wind, constantly looking down, since they locate fish by sight. When a suitable shoal of fish is located, they dive with wings half-closed and angled backwards; as they enter the water the wings are stretched back behind the bird, but are not folded. The diving is best seen from a trawler, for when the net is hauled there is a time when it lies alongside the ship; during this period it appears to be raining gannets as they dive at the net and at any small fish escaping through the meshes. Surprisingly few get caught in the net; in our own experience during two weeks' trawling, only one gannet was enmeshed and drowned.

The gannet is a large bird with a wing span of about 6 feet. The wings are long and narrow, suited for fast flight, and consequently landing and take-off are not easily accomplished. From the water they have to head into wind and splash along for some way before they have enough speed to rise. From the colony they usually launch themselves from the cliff-edge, but on large flat-topped sites the birds well away from the edge may have to make an ungainly, crashing take-off run.

The gannet, being a fast-flying bird, has to undergo vigorous braking movements as it lands. Speed is lost by sweeping the wings forward in a 'back-pedalling' movement, the spread feathers giving the wing maximum resistance to the air; further braking is gained by the extended feet and fanned tail feathers. All this effort is accompanied by a rhythmically uttered loud, harsh, guttural cry, a note only used at other times in anger or during the excitement of display.

Gannets do not achieve adult plumage for some years, passing through a number of speckled or pied immature plumages; the photograph shows a young bird, in its third summer, coming in to land on the edge of a gannetry. The gannet has a high aspect ratio wing, but this ratio can be lowered by feather movements when flying slowly. The wings of this landing gannet are broader than those of gannets flying faster. Since a long, narrow (high aspect ratio) wing has a high stalling speed there are marked anti-stalling devices in the form of wing-tip slots and a large bastard wing. Extra lift at low speed is achieved by a large angle of attack. The wings are being swept forward in a braking action, supplemented by the extra drag of the flexed and spread tail and the lowered feet.

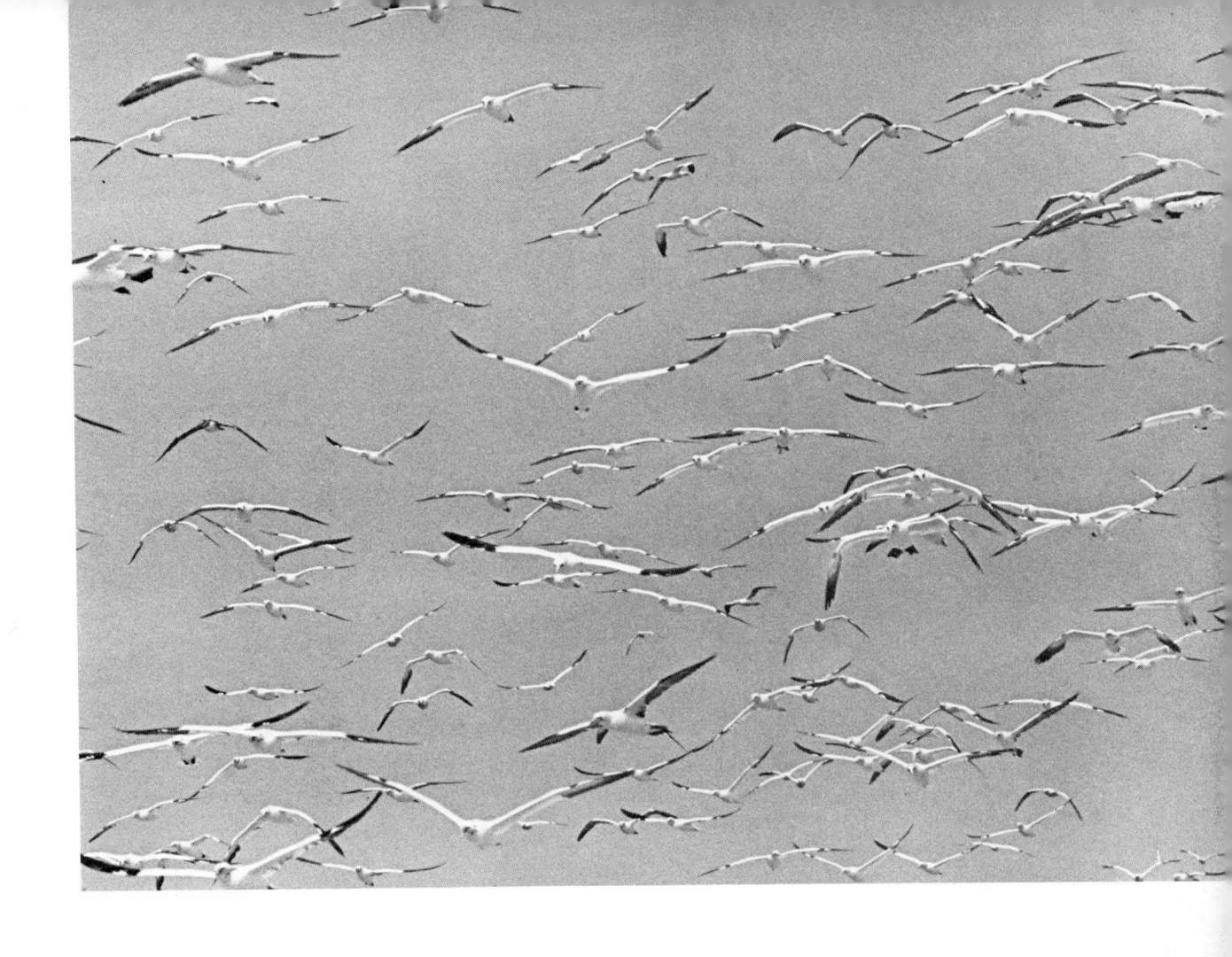

The air over a gannetry is a complex kaleidoscope of brilliant white shapes. Hundreds of gannets are circling, with individuals constantly dropping noisily to their nests. The gannets pass over the colony facing into the wind, so as to reduce their landing speed, and then sweep in a wide circle downwind over the sea.

Flying at medium speed over a gannetry. Drag is reduced at these speeds by the wing feathers being held so as to give a long, narrow wing, and by a low angle of attack. The birds are flying well above their stalling speed, and the bastard wing is held so as to merge with the leading edge of the wing.

This gannet clearly illustrates the feather arrangement of the wings. There are definite tracts of feathers other than the primaries and secondaries that have already been described. The bases of the primaries and secondaries are covered by the under wing coverts, close to the body there are the axillaries, and the wing linings are small feathers under the leading edge. These feathers serve to give a smooth aerofoil shape to the wing.

Two immature gannets just about to commence landing.

A gannet in level flight with the wings on the upstroke and the feet tucked away where they will not create turbulence or drag.

The difference between the upstroke and downstroke of the wings is clearly seen in the lower photograph. The wings of the near bird are being moved upward, slightly bent, with the outer part of the wing drooping; the centre bird in the distance is moving the wings downward, the primaries bent up by air resistance.

SUMMER SHORES AND ISLANDS

The final braking for landing is just about to commence; the wings of this gannet will be swept forward, the feet lowered still further, and the tail, already somewhat flexed, will be spread.

A gannet banking at speed as it circles over the gannetry.

SUMMER SHORES AND ISLANDS

When departing from a gannetry a boat sails for some time beneath a canopy of these magnificent birds.

Shags

SUMMER SHORES AND ISLANDS

The shag is also known as the green cormorant, and indeed there is a great similarity in the overall appearance of shags and common cormorants. The shag is more strictly marine than the cormorant, and, whereas cormorants are found on both muddy and rocky coasts, the shag breeds where the coastline is composed of rock.

The shag is a strong flier, quite fast, travelling through the air in a direct manner with its long neck stretched out in front of its steadily beating wings.

Like many other fast-flying birds it has to expend considerable energy in braking to land, often assisting this if the nest is on a high ledge by approaching fairly low and sweeping up to the landing site. Taking off from a ledge is easy, the shag simply launching into the air and gaining speed as it dives, but it is more difficult from the surface of the water; they splash over the water for some way before rising, leaving a regular pattern of disturbance on the water with every wing beat.

Both shags and cormorants dive for fish, and subsequently can be seen standing with their wings held out to dry, occasionally flapping them to speed the process. It is probable that this behaviour is necessary as their wings are less well waterproofed than those of most diving birds.

Shags are sleek birds on the ground or in level flight, but when braking hard as they come in to land they are immensely untidy. Wings are vigorously flapped, the tail is spread to an almost excessive extent, while the large feet frantically add their assistance. The slotting effect of the spread primaries and the bastard wing can be clearly seen in these photographs.

SUMMER SHORES AND ISLANDS

This shag shows the feather arrangement of the upper wing; the primaries and secondaries are overlain at their bases by the upper wing coverts, arranged in two main series, and in the angle formed between body and wing are the scapulars.

Kittiwakes

SUMMER SHORES AND ISLANDS

The kittiwake is more strictly marine in its habits than any other British gull. Apart from during the breeding season it is an unusual sight to see a kittiwake on land. When they come to land to breed they choose to nest on some of the sheerest cliffs. The nests are built in colonies, and are quite substantial structures built with amazing skill, often on only the merest suggestion of a ledge.

On the wing the kittiwake is lithe and graceful. The wing beats are faster than those of most gulls, and the swift flight gives a marked impression of lightness and buoyancy.

The name kittiwake is an attempt to write down the normal call of the bird; hundreds of these calls echoing from the walls of a rocky cove produce a sound whose wildness lingers as a haunting memory long after summer. At frequent intervals a focus of intensity in this sound can be heard as a pair display, or as one of the intermittent squabbles breaks out. These disputes result in aerial chases from the cliff-face, usually of short duration, although occasionally we have seen a bird forced on to the water, and long battles fought between three birds on one nest.

Incredible flying skill is exhibited during the aerial arguments, and the kittiwake's mastery of the air is shown well on days of high wind, when they will play with the wind, exploiting it or overcoming it as the need arises.

One of the easiest ways to distinguish the kittiwake from other gulls in flight is by the black wing tip unrelieved by the white spots or 'mirrors' of most other gulls with black on the wing tip.

SUMMER SHORES AND ISLANDS

When disputes break out in the colony, an aggrieved bird will often turn in a tight circle to rejoin the argument, loudly protesting all the time. This kittiwake is in fact executing such a turn, banking steeply, its bill wide open uttering a stream of abuse.

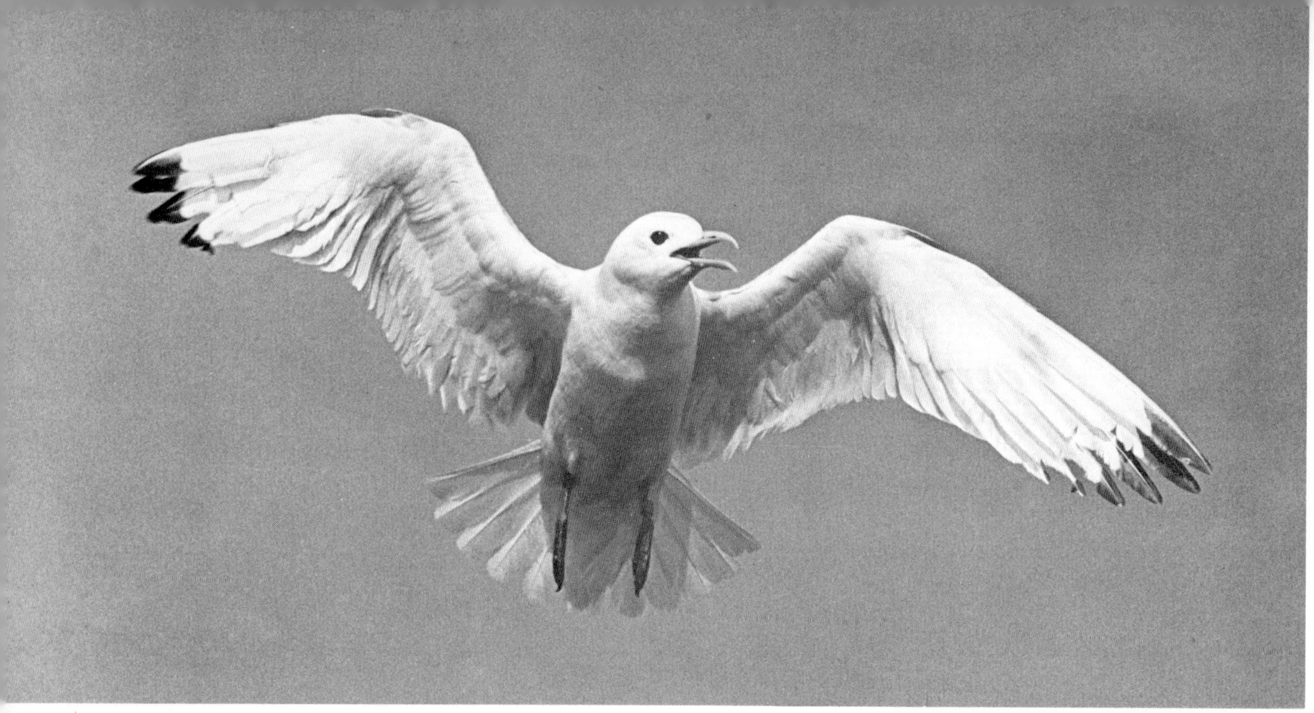

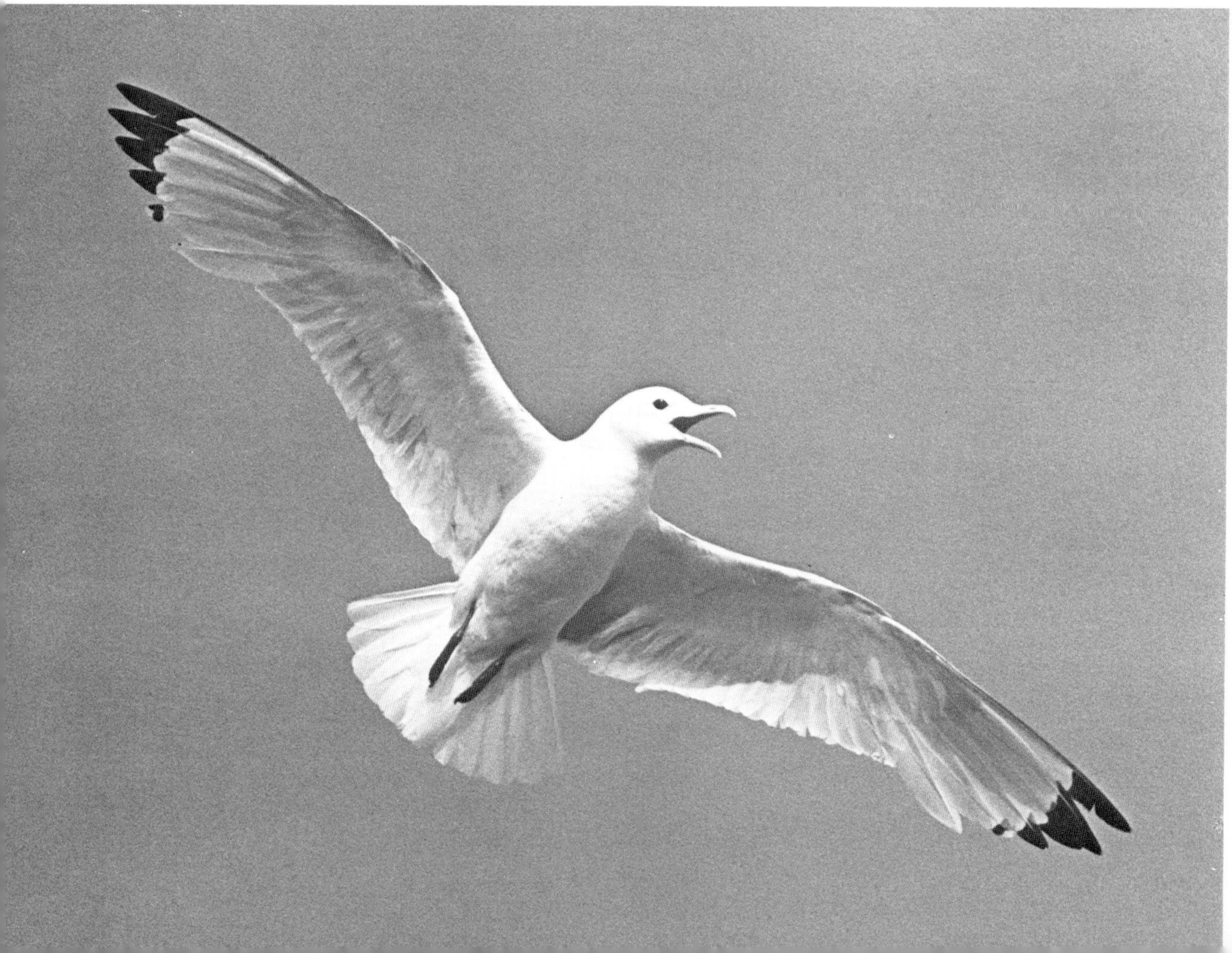

Kittiwakes display great flying mastery as they wheel in front of the cliff colony. Especially impressive is their performance in high, gusty winds, when they will rise on updraughts or descend by spilling the air from their wings with practised ease.

The almost black legs of the kittiwake are very conspicuous when they are used as steering or braking aids in flight.

SUMMER SHORES AND ISLANDS

The kittiwakes below and opposite have their feet lowered, ready to spread for full aerodynamic effect.

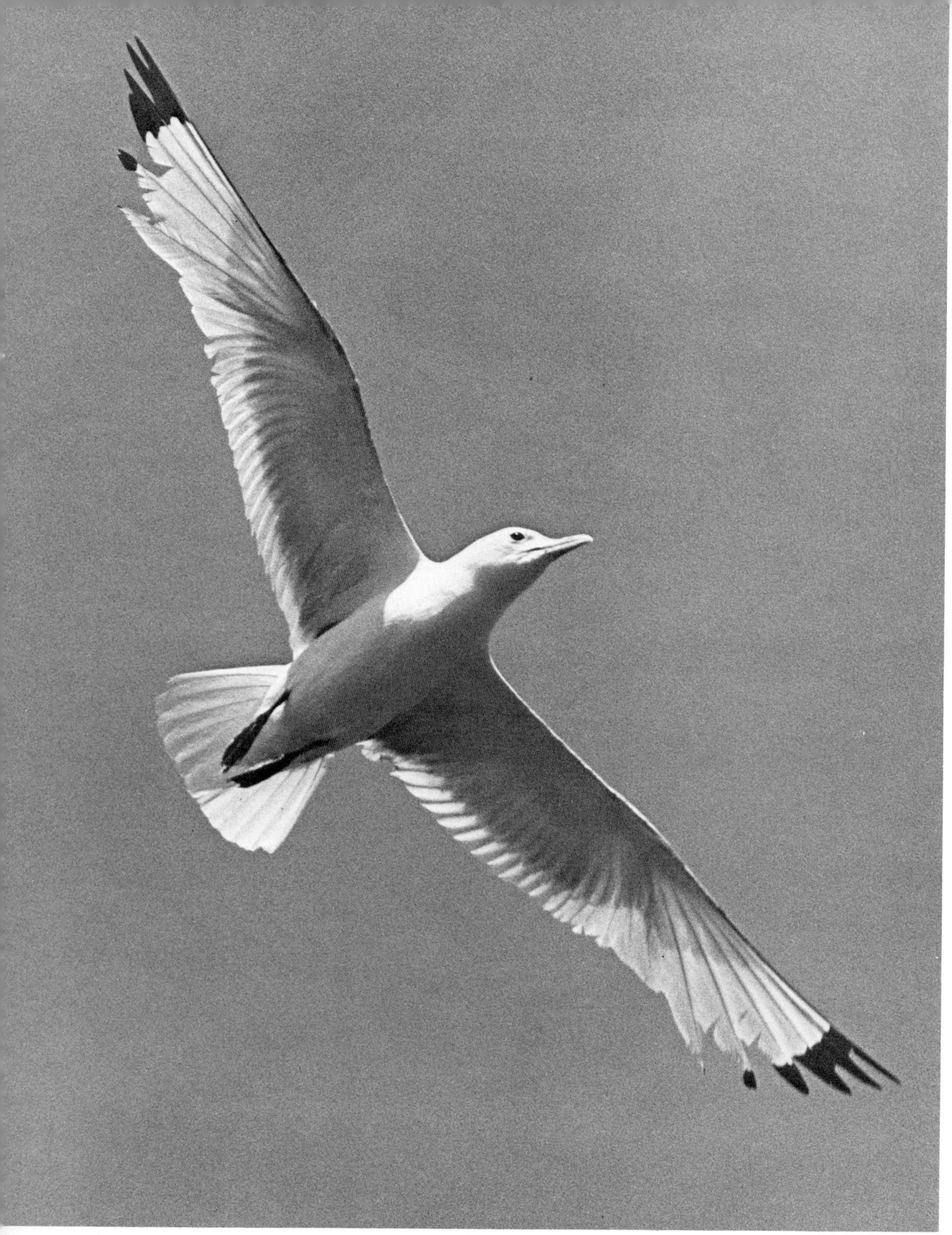

SUMMER SHORES AND ISLANDS

The wings of the kittiwake are more translucent than the wings of heavier scavenging gulls. Light can be seen passing through the primaries and secondaries, but there is a definite line where the extra thickness of the coverts makes the wing opaque.

Flocks of kittiwakes can often be found resting on flat rocks or sloping broad ledges near the colony. These birds are very wary, rising in the air together at any disturbance.

Lesser Black-Backed Gulls and Herring Gulls

These gulls are scavenging gulls, and are not so strictly marine as the kittiwake. They usually nest in colonies, not on the precipitous cliff-ledges preferred by kittiwakes, but on wider, often grassy ledges, or the flat tops of grassy islands, often amongst bracken. Both will occasionally nest at some distance from the sea, by fresh water or on bogs, but this is less frequently the habit of the herring gull.

The flight is purposeful and powerful, and has a versatility suitable to scavenging birds. They have a large wing area and can soar on the upcurrents of air created by cliffs as they patrol, ever watchful for a meal. The powerful level flight is of a different type from the landing and take-off flight. A different type of wing action allows them to drop precisely and easily to the ground, and to rise steeply from flat ground or the water. This same wing action allows them to hover momentarily as they pluck something from the ground or sea. In a bird with such a large relative wing area these energetic actions can only be maintained for a short while without fatigue, but the ordinary level flight uses only the powerful, larger set of flight muscles and can be undertaken for long periods of time.

Any intruder to the nesting colony is discouraged in the most effective manner, being repeatedly subjected to spectacular dives. The gull will climb to a fair height, and then swoop down in a steep glide passing at considerable speed within inches of the unwelcome visitor with a loud rush of wings. Although the feet, sometimes both, sometimes one, are extended as the gull passes over the intruder, it is seldom that an actual strike is made. The performance is, nevertheless, most intimidating, and even if it does not completely succeed in frightening away the intruder, it will serve to distract attention from the camouflaged chicks.

The wings of the lesser black-backed gull are much heavier and more opaque than those of the kittiwake. The 'mirrors', white patches towards the tips of the wings, can be seen.

SUMMER SHORES AND ISLANDS

Two attitudes of lesser black-backed gulls during the diving attack on an intruder.

Before making its intimidating dive the lesser black-backed gull climbs to some height, hovers momentarily, and then glides steeply down, gaining considerable speed.

SUMMER SHORES AND ISLANDS

More attacking dives by lesser black-backed gulls.

Herring gulls coming in to land on a rock-perch at the top of a cliff. The wings are flapped backwards and forwards energetically as the bird slowly drops the last foot or two to the ground.

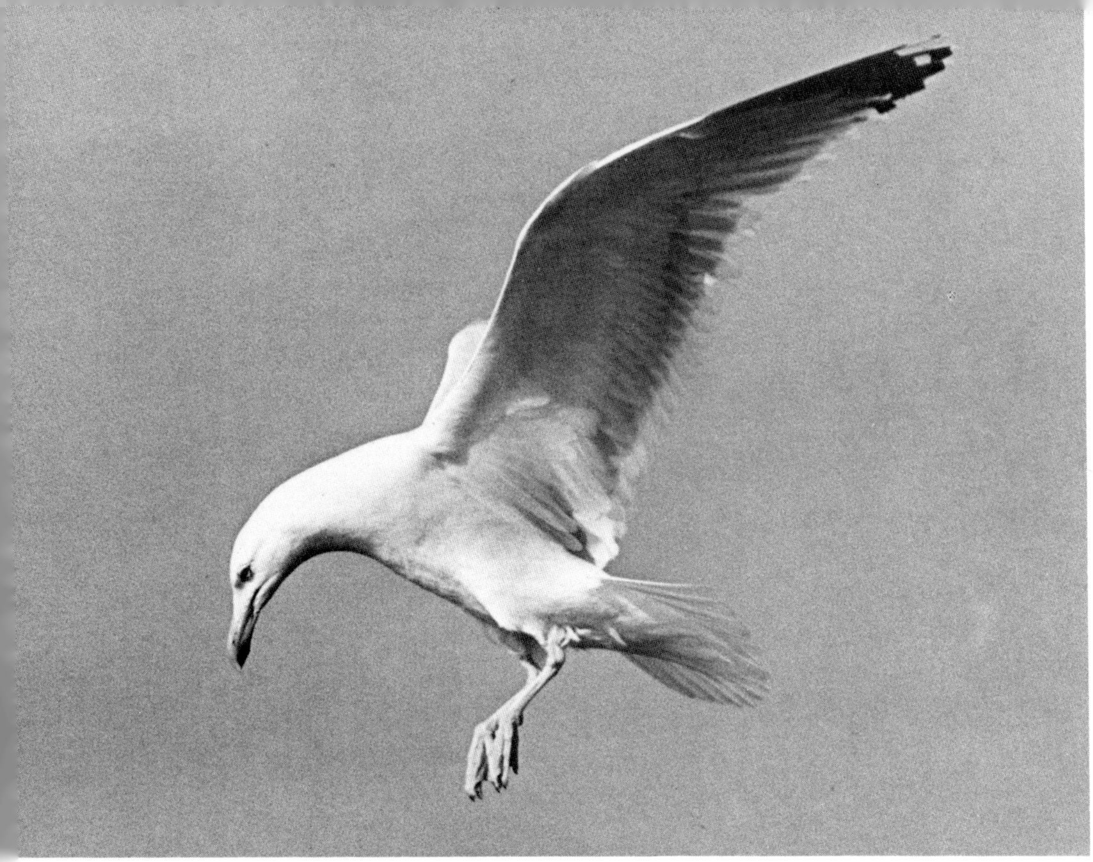

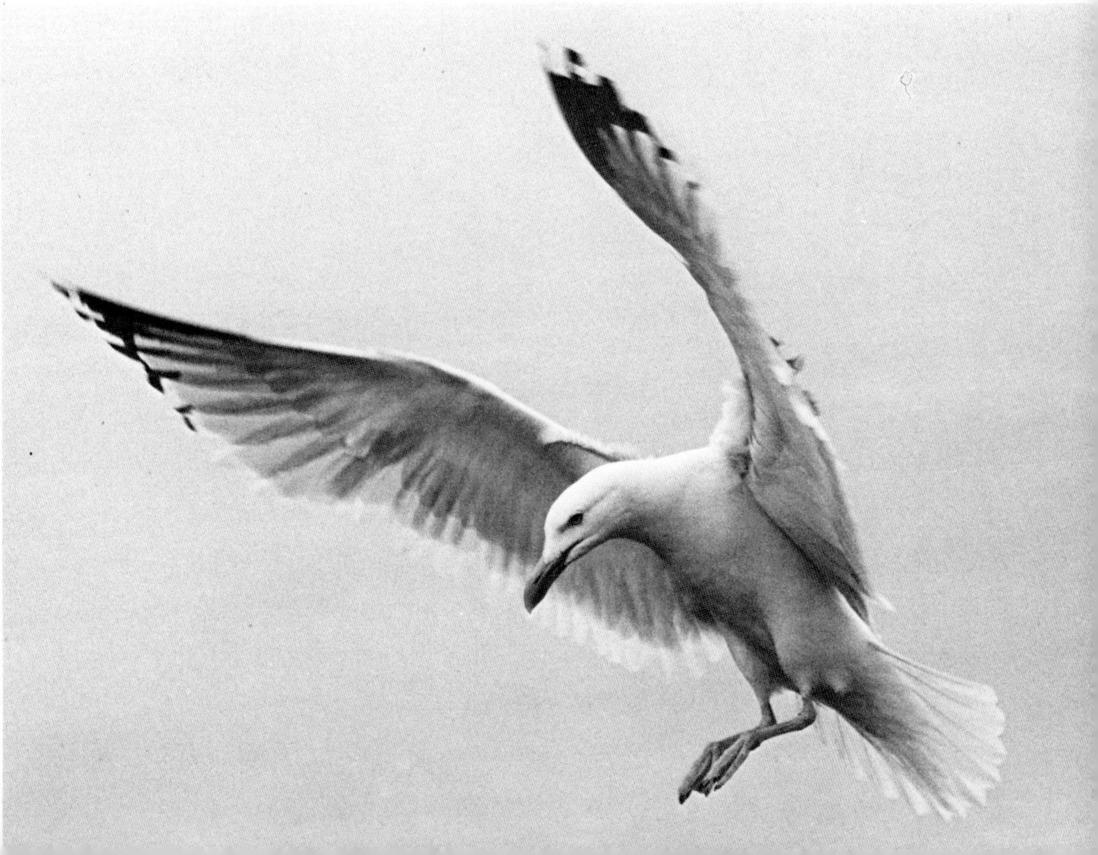

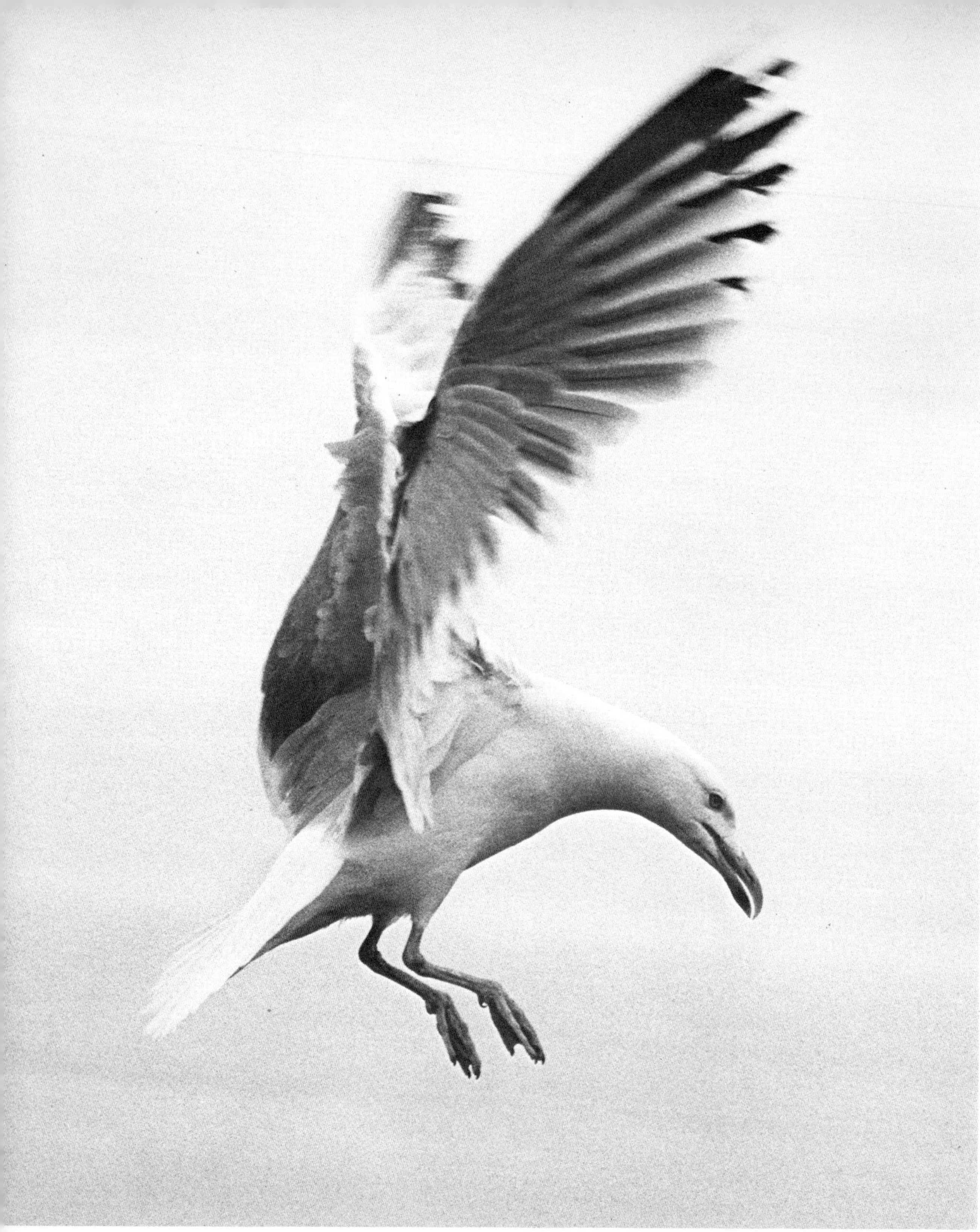

SUMMER SHORES AND ISLANDS

A herring gull hovers momentarily to inspect a landing-site. This ability to pause in the air is a great asset to a scavenging bird.

SUMMER SHORES AND ISLANDS

A mixed flock of gulls rising from an estuarine feeding and resting ground. The near-by cliffs and islands are empty of breeding birds, and there is a brief interval before the silence left by their departure is filled with the roar of a sea angered by autumn gales.

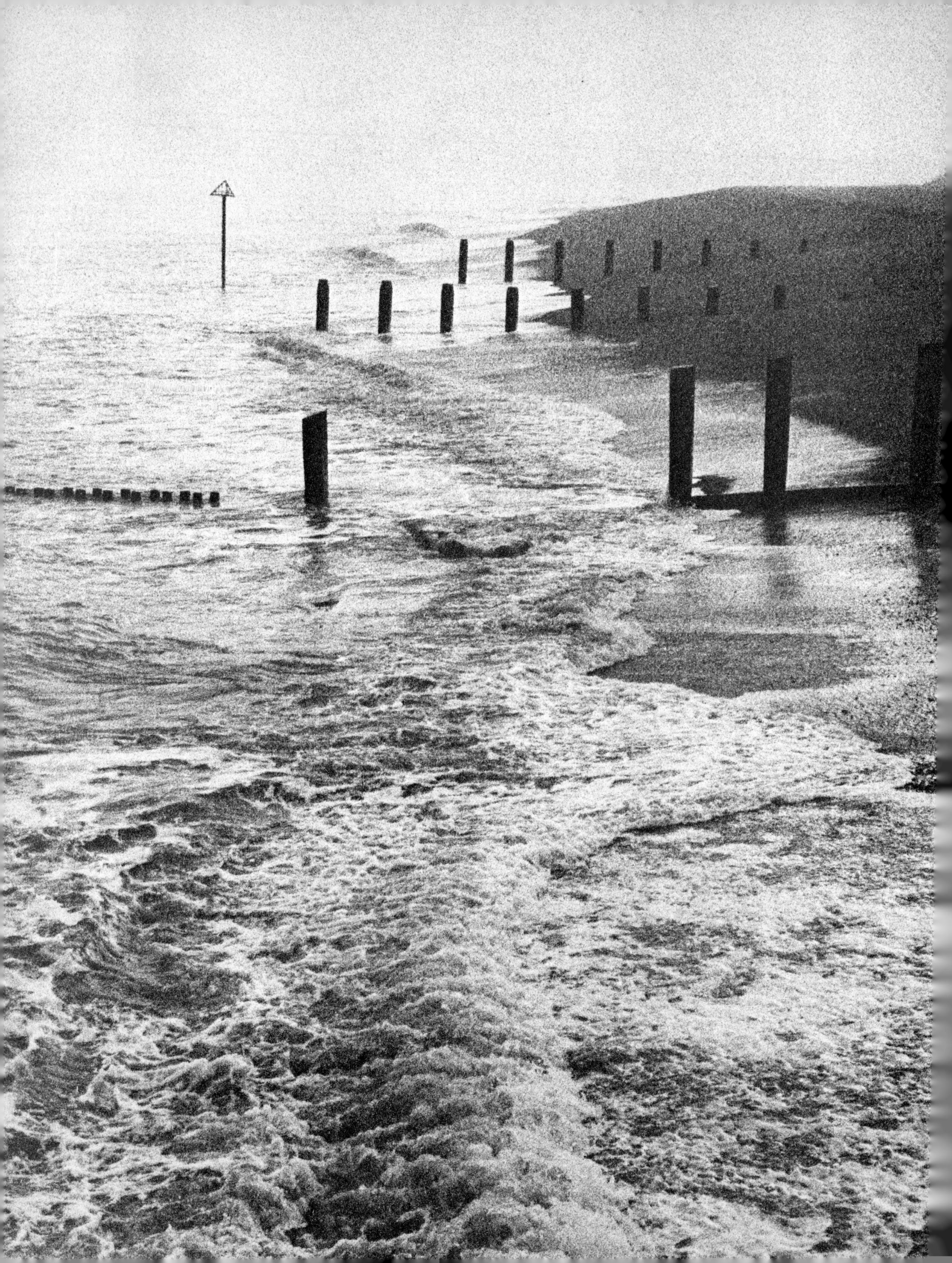

Winter Shores and Marshes

The cliff-ledges are now empty of their summer populations of sea birds, but when the summer is spent, fresh multitudes of birds come flying on the autumn air, congregating on flat shores, estuaries and marshes. The interlacing traffic of wings is now replaced by fast-moving patterns of birds, waders, geese, swans and ducks. The flat and subtle hues of the marshlands form a delicate setting for the bold geometry of these wheeling flocks.

Some of these birds can be seen in smaller numbers throughout the year; in winter their numbers are dramatically reinforced by migratory members of their species, and by birds which visit Britain only during winter. These winter visitors have come from higher latitudes, from Scandinavia, Siberia, Greenland, Iceland and Spitzbergen.

The daily movement of these flocks is governed by tide and time as they fly to and from their feeding-grounds, and this is when they are best seen in flight. Many of these marsh and shore birds are shy and wary, and little will be seen if they are carelessly approached. Caution and patience are necessary, but both are amply rewarded when a skein of geese fly low overhead filling the air with their wild sound, or a flock of waders leave the edge of a rising tide, stirring the icy air with their slender wings.

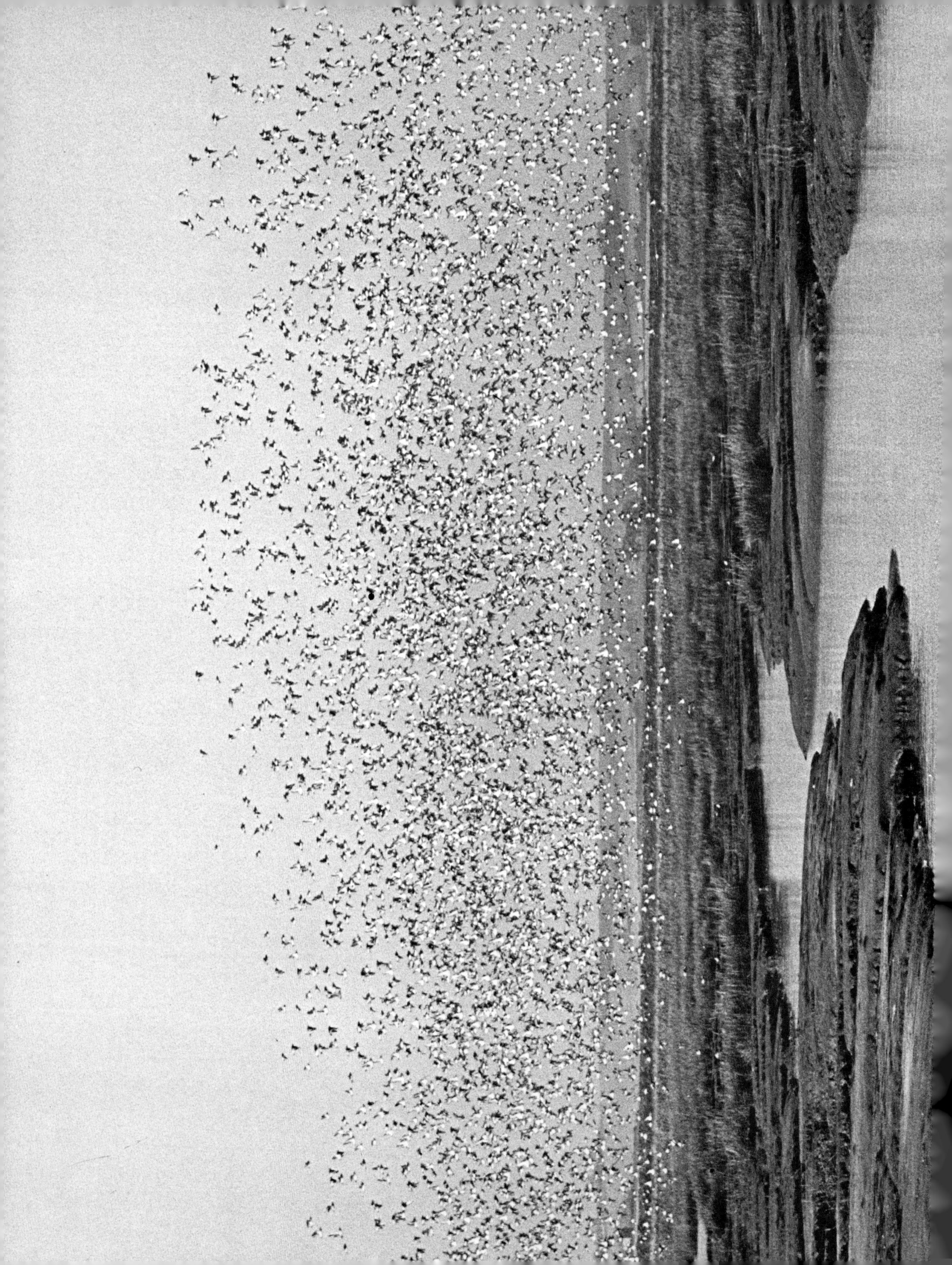

On a rising tide the waders that have been feeding on the expanses of mud become more and more concentrated in smaller areas as the water advances. The birds then form cohesive flocks and fly in large groups to the last uncovered patches of mud. The picture above shows one of these flocks tumbling out of the sky, spilling the air from their wings, as they drop to a feeding ground high up the shore as the tide advances. Eventually all the intertidal mud is covered and the waders take off in vast flocks; part of one of these flocks is seen rising from the high shore in the photograph opposite.

Some wading birds display bold and distinctive markings when flying; these can be seen in the oystercatchers above. Various positions of the very full wing beat used during take off can be seen.

The turnstones above display a bold plumage pattern in flight; in the lower picture, as well as oystercatchers, are some of a more discreetly marked species, the knot.

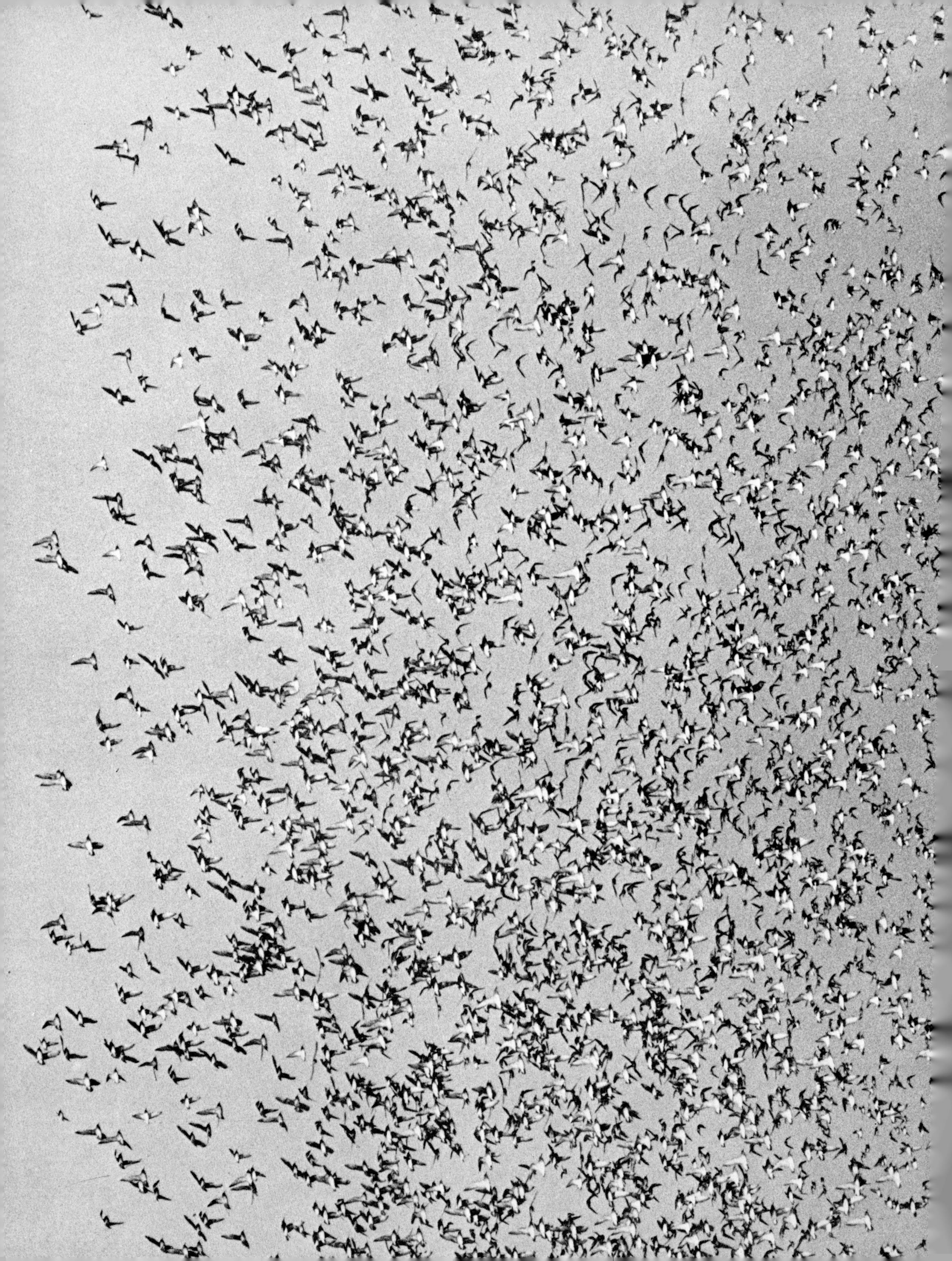

On the opposite page a dense flock of knots are climbing from the shore as the tide covers their feeding grounds. A group of bar-tailed godwits form a pattern of silhouettes against a drifting winter sky in the picture above.

The rising tide causes oystercatchers to leave the shore. When a large number of these comparatively heavy waders take off simultaneously, the sound of their wings can be heard for a considerable distance.

WINTER SHORES AND MARSHES

A flock of oystercatchers flying along the shore, over the flood tide.

Oystercatchers rise as the tide advances over the muddy shore.

WINTER SHORES AND MARSHES

More activity as the marshes flood, and the oystercatchers are again obliged to leave.

Knots breed in the Arctic in summer, migrating to more southerly latitudes in winter. The wings are of high aspect ratio, suited to fast flight.

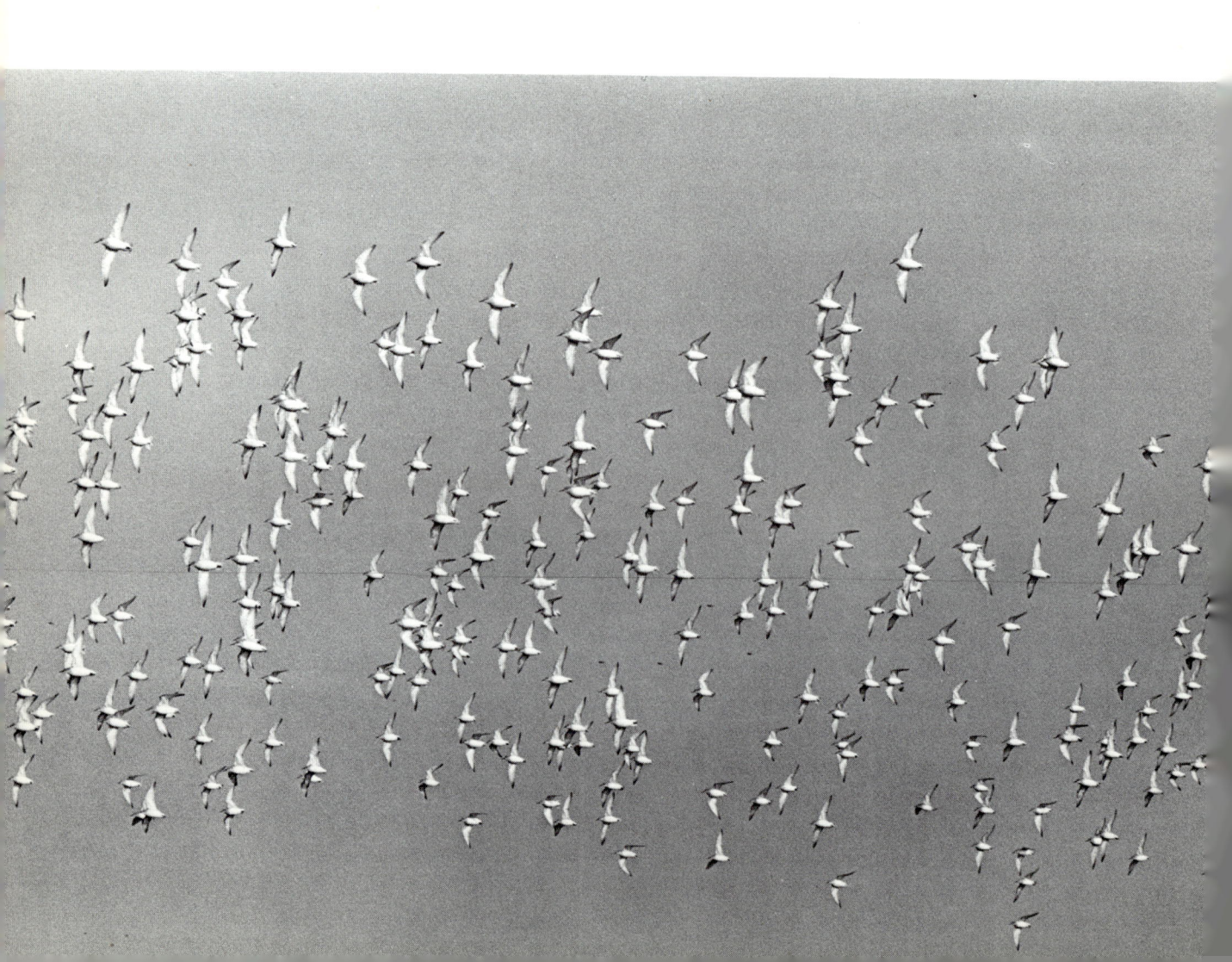

Knots form some of the densest wader flocks, and when flying these form constantly changing patterns, dark one second, then turning in complete unison, their pale undersides flashing in the pale wintry sunlight.

The mute swan is found on ponds and lakes and is the semi-wild swan familiar in Britain. In the winter some assemble on marshy coasts, where in certain places the numbers may reach several hundred. The wing beat of the mute swan produces a characteristic singing sound, audible for some distance.

In Britain, during winter, the swan population is reinforced by numbers of two species of wild swans; one of these is Bewick's swan. This swan winters mainly in England, often away from coasts, but also on marshy shores.

The other wild swan to visit British coasts in winter is the whooper swan. Although a few of these remain in Scotland in summer, and occasionally nest, these are joined by two to three thousand immigrants, probably from Iceland, each autumn. Most of the whooper swans remain in Scotland and northern England, reaching southern England in much smaller numbers. Neither whooper nor Bewick's swans' wings make the singing noise of the mute swan.

A group of Bewick's swans photographed through the bare branches of winter trees. These are part of a flock that, thanks to skilful management, have visited the Wildfowl Trust in Gloucestershire in increasing numbers over the past few winters.

A group of mute swans silhouetted against the pale winter sky as they fly over the shore. The birds with their wings on the downstroke show how the tips of the primaries are bent upwards by the reaction of the air.

WINTER SHORES AND MARSHES

The wing beat of all swans is slow and powerful. These Bewick's swans are rising over the trees as they fly to a feeding ground near the edge of an estuary.

Whooper swans coming in to land on an expanse of wet mud, appearing very dark as they are silhouetted against the glittering light which is streaming through the translucent flight feathers.

Bewick's swans in level flight.

A group of mute swans fly past as they move to another part of the shore.

WINTER SHORES AND MARSHES

The wings of Bewick's swans combine with out-of-focus trees to form contrasting patterns against the grey winter sky.

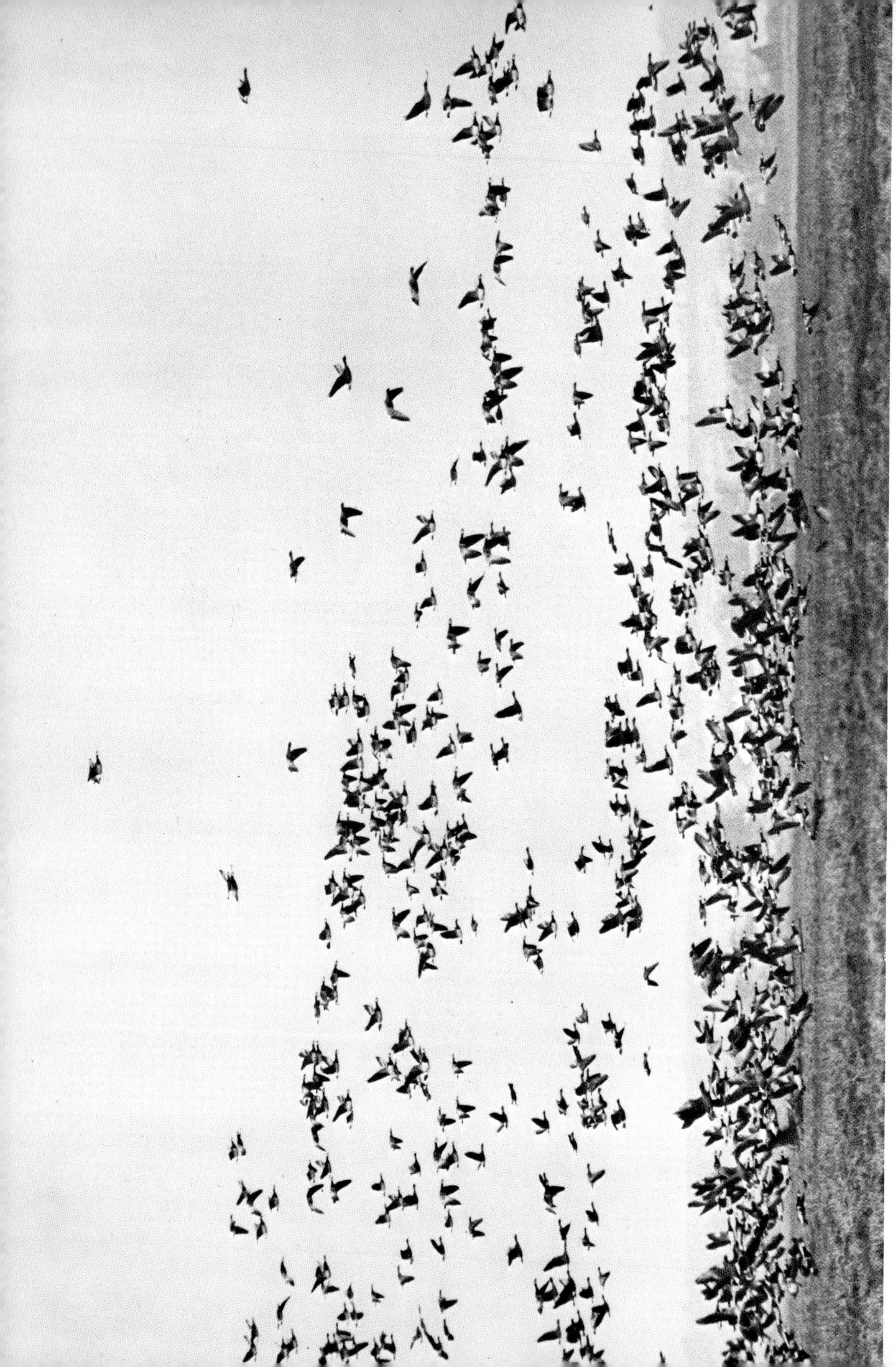

The barnacle goose is handsomely marked in black, grey and white, and in winter visits Scottish western shores and islands in large numbers. Mingled with them in these pictures are a few pink-footed geese which feed on the same salt marshes as well as further inland.

Barnacle geese and a few pink-footed geese rising shortly after take off. These birds are moving to another feeding area on the same salt marsh, and if flying only a short distance geese do not adopt their characteristic formations of vees or trailing lines; nevertheless, there is a pleasing regularity in the spacing of the birds.

WINTER SHORES AND MARSHES

The feeding habits of the brent goose confine it more strictly to shores than the other geese visiting Britain in winter. A line of these small, dark geese are shown flying along a creek, silhouetted against the sparkling wet mud.

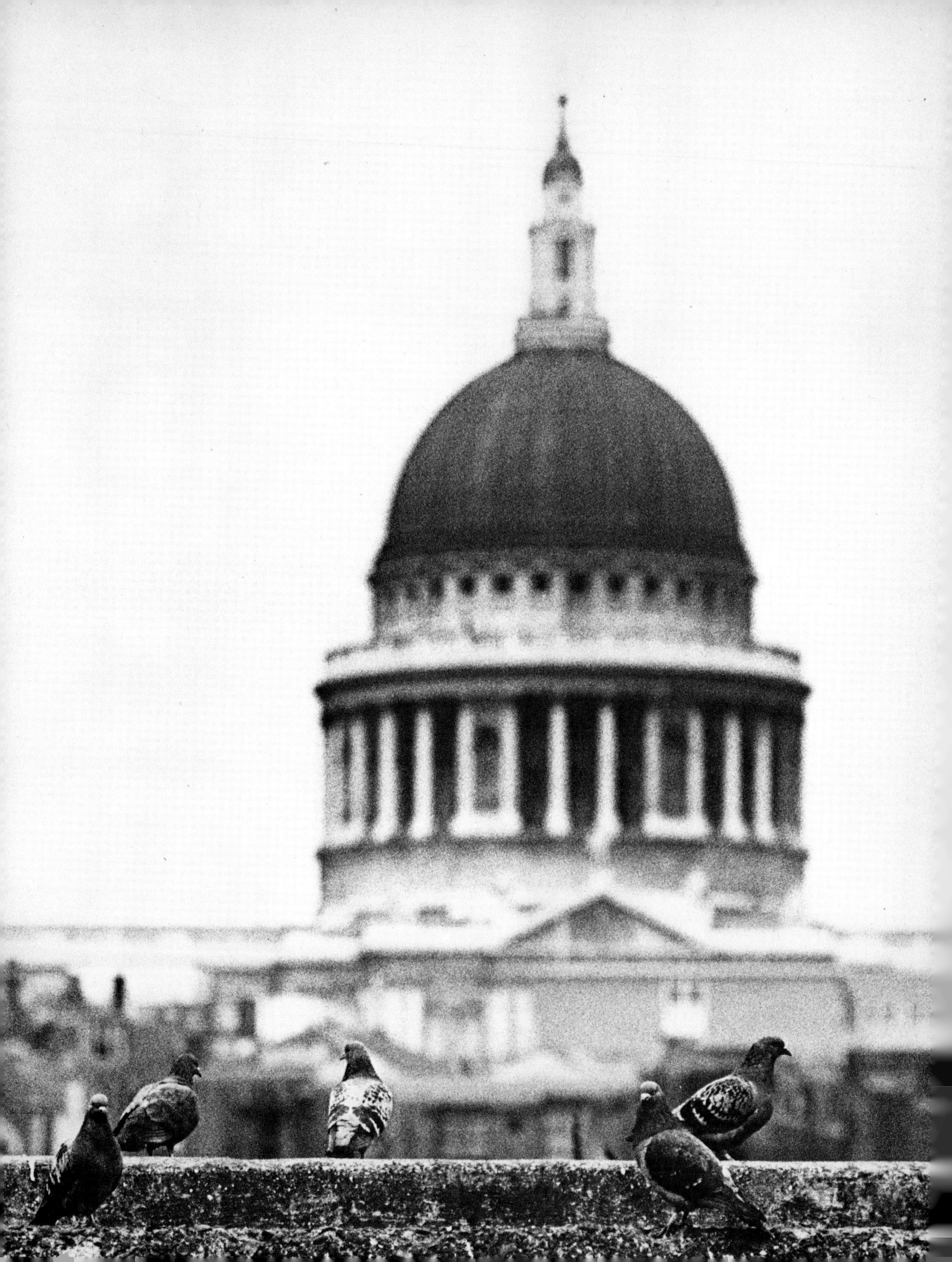

Winter in the City

The birds of the flat and marshy shores are all too easily disturbed by man, and, in an increasingly populated and industrialised country, reserves and refuges are often necessary for their protection. Other birds, more adaptable in their habits, not only tolerate the presence of man, but live in great numbers in the city.

These birds are at no time more evident than in the winter, when large starling flocks will roost on buildings, and black-headed gulls patrol the parks and riverside, joining the ever-present pigeons and sparrows. The extreme tameness of city-dwelling birds makes the appreciation of bird flight an easy and leisurely occupation.

The pigeon has been used extensively in studies on bird flight. Apart from the ease with which it is bred and kept in captivity, it is also a suitable subject for basic studies, having no extreme modifications of wing shape or flying behaviour. The wings of this bird are on the downstroke as it approaches a landing-perch; the speed has been reduced and the bastard wings can be seen raised from the leading edges.

A London pigeon, just beginning the downstroke of the wings, flys in front of the spire of St. Martin-in-the-Fields.

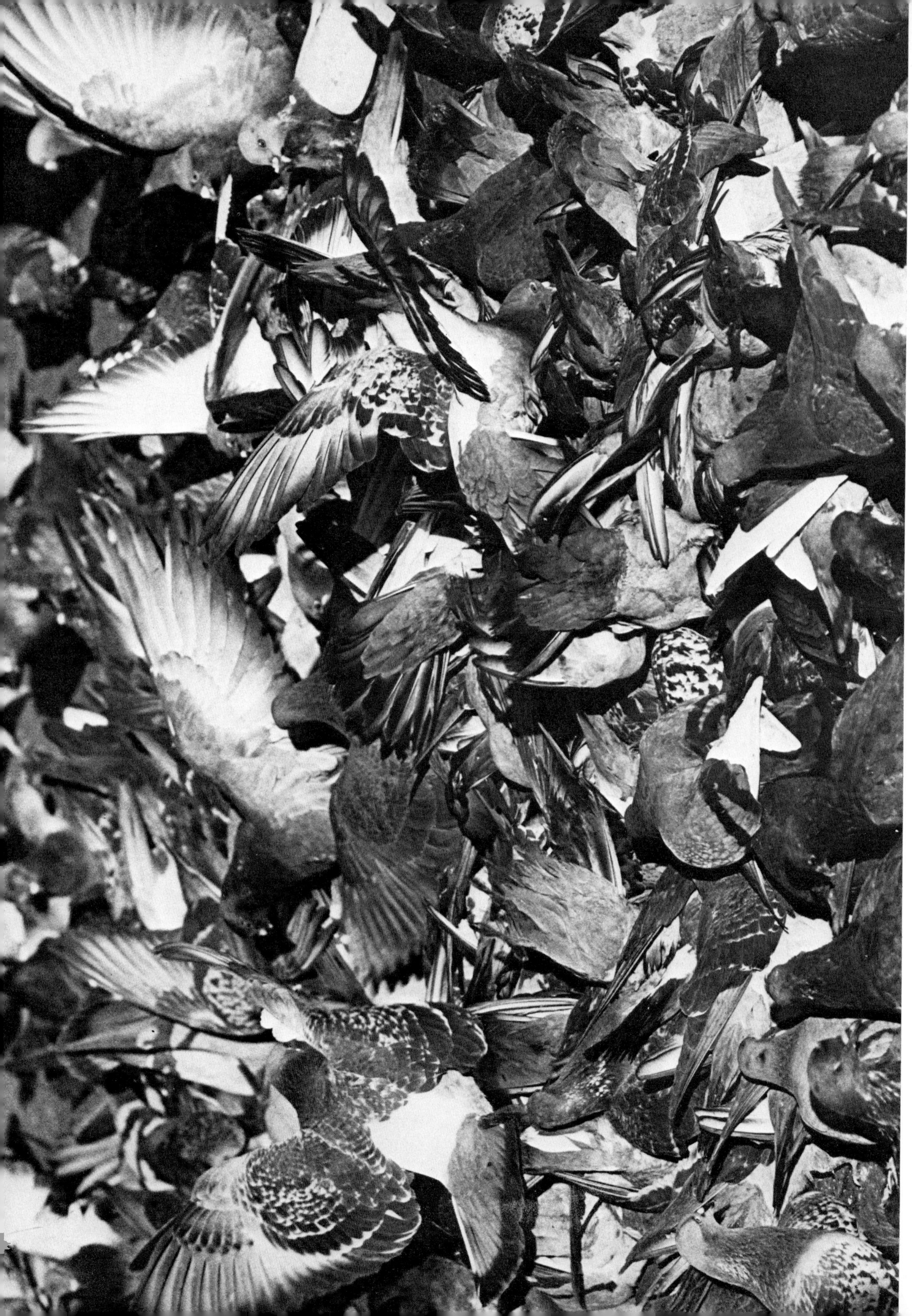

WINTER IN THE CITY

The pigeons of Trafalgar Square provide one of the tourist attractions of London, and their presence in London has been recorded for over six hundred years. When grain is thrown down a large number of pigeons descend, and the pavement is covered by a mass of flapping wings. This sort of mêlée has no doubt contributed to the dishevelled appearance of the bird below.

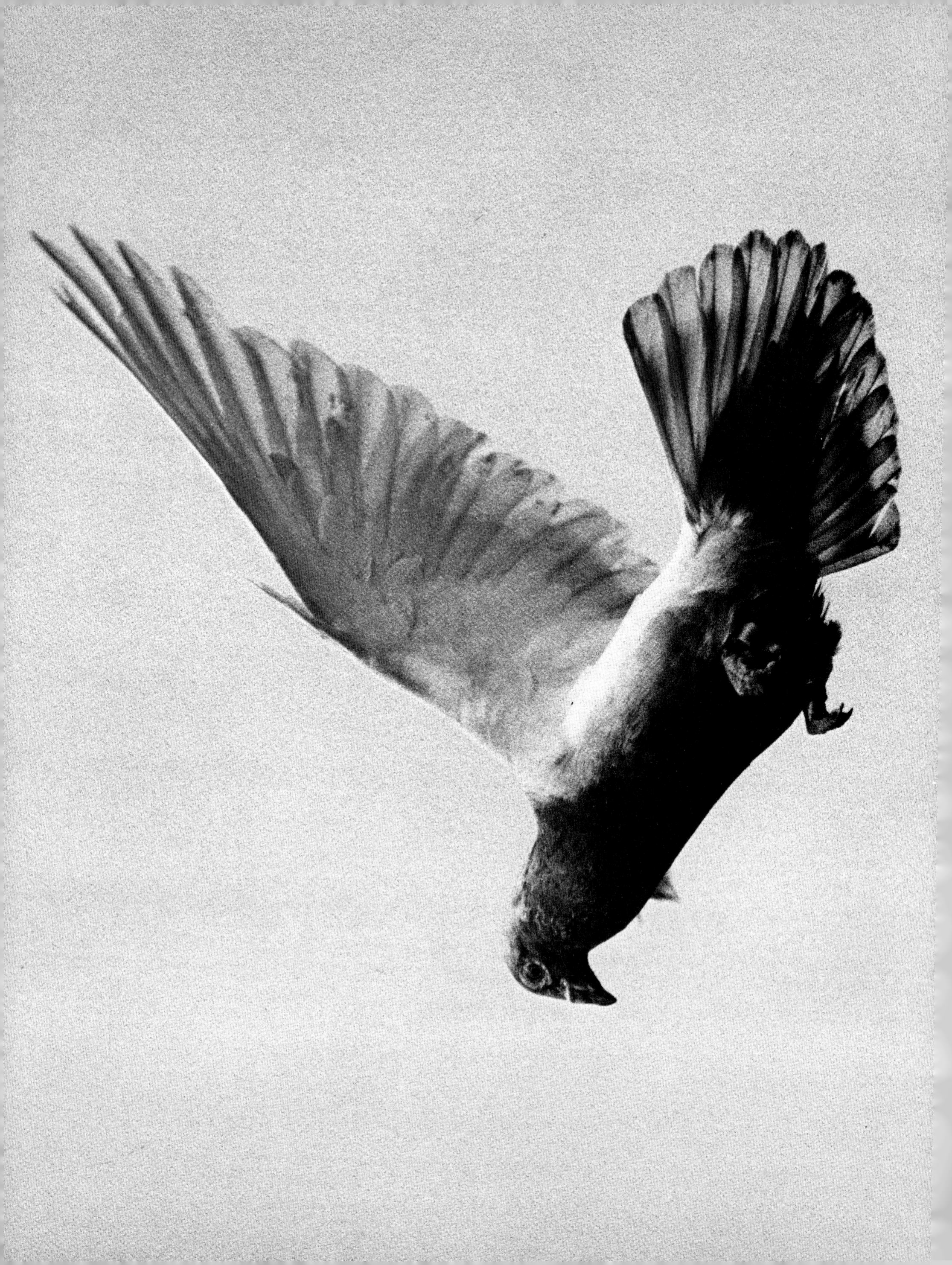

Another familiar urban sight, particularly in more northern English cities and towns, are the pigeons specially kept for racing. The bird in the photograph above is employing the type of wing movement used for short periods at take off and landing.

The Thames-side wharves, with abundant spilled grain, provide food for large numbers of pigeons, and, in winter, many black-headed gulls. Various positions of the wing during its action in landing and taking off can be seen in these two photographs, the wing action of gulls and pigeons being basically similar (the position of one of the pigeons towards the right-hand side of the picture above is almost exactly echoed by a gull on the left of the photograph opposite).

PRIVATE
NO PARKING

The wings of this black-headed gull are in the flexed position of the upstroke. In their winter plumage these gulls do not have the nearly black cap to the head, but merely a dark spot behind each eye.

WINTER IN THE CITY

The London parks hold large numbers of black-headed gulls in winter, and any scraps of food thrown down will immediately attract a crowd of these scavengers.

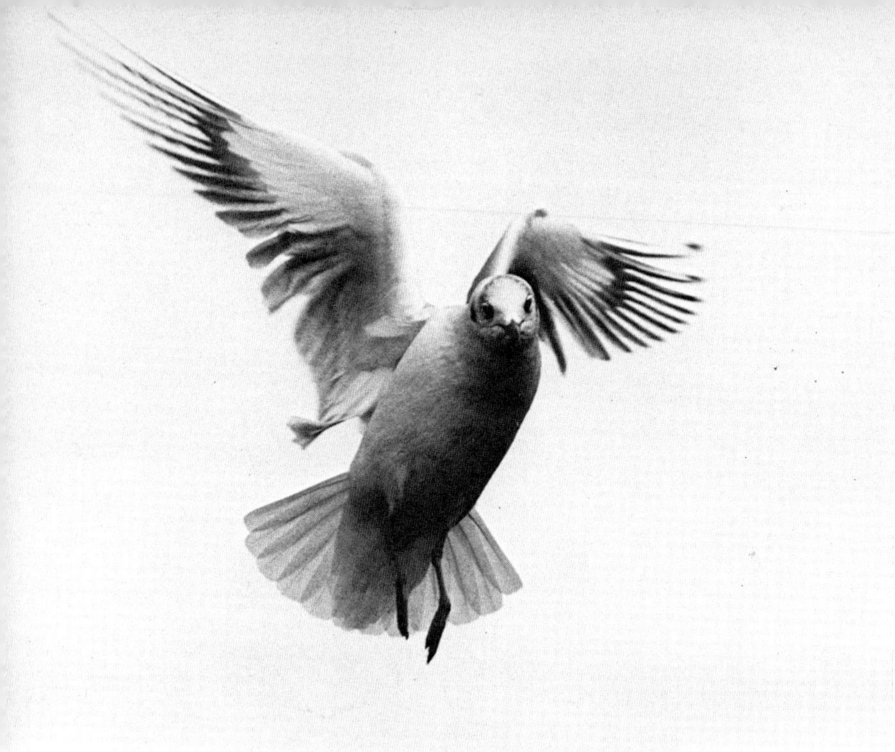

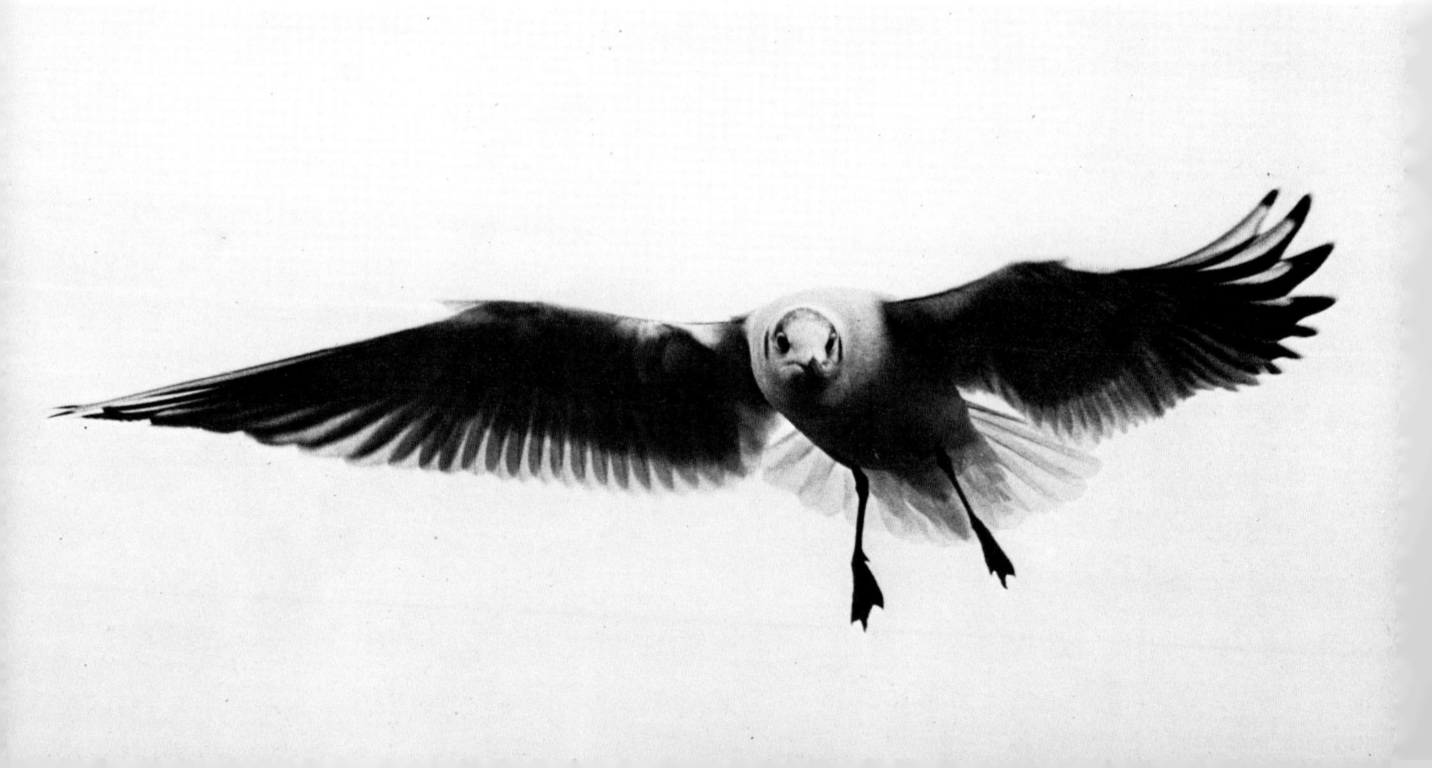

Black-headed gulls have been regular winter visitors to London since 1895, and they have subsequently become increasingly numerous. They perform the role of scavengers, and as such replace the kites which left London in the mid eighteenth century. One of the factors contributing to the success of black-headed gulls as scavengers in cities is their agility on the wing. They can hover momentarily, change direction sharply, and manœuvre with great dexterity. A first-rate display of aerobatics takes place over the head of anyone throwing bread in the air for these gulls.

One of the inevitable consequences of urban communities are the rubbish-tips on the outskirts of towns or cities. The fresh garbage provides plenty of picking for gulls, and attention is often drawn to these sites by the white clouds of hundreds, or even thousands, of gulls that rise and descend as the bulldozers level and consolidate the material deposited by the dustcarts.

During the winter, outside their breeding season, the black-headed gulls form the most numerous inhabitants of the rubbish-tips. Some of these hover, almost stationary on the wind, searching for food; others strut about turning over the debris on the ground, whilst there are also usually large flocks of well-fed resting birds, forming white patches in the near-by fields.

The ubiquitous starling is a familiar bird of towns. At the end of the breeding season these birds begin roosting socially in very large numbers, and in some cities these roosts are situated on buildings. London holds some of the largest urban roosts in Britain, although this habit was only established there at approximately the beginning of this century. Starlings may fly for 20 or 30 miles from their roosts to their daytime feeding grounds, and groups of the swift-flying, small, dark birds are a familiar sight in the late afternoon sky of the suburbs, as they hurry back to their roost. As more and more of these parties arrive at the roost the level of noise increases, and the air is often filled with spectacular patterns of flying starlings before they finally settle for the night.

WINTER IN THE CITY

During the day the starlings fly from the roosts to feed, and parties of various sizes can be seen in a wide variety of places. In the urban environment some of the largest flocks are to be found feeding with the gulls on the rubbish-tips. They fly in densely packed flocks from one part of the tip to another, seeking fresh areas to explore for food.

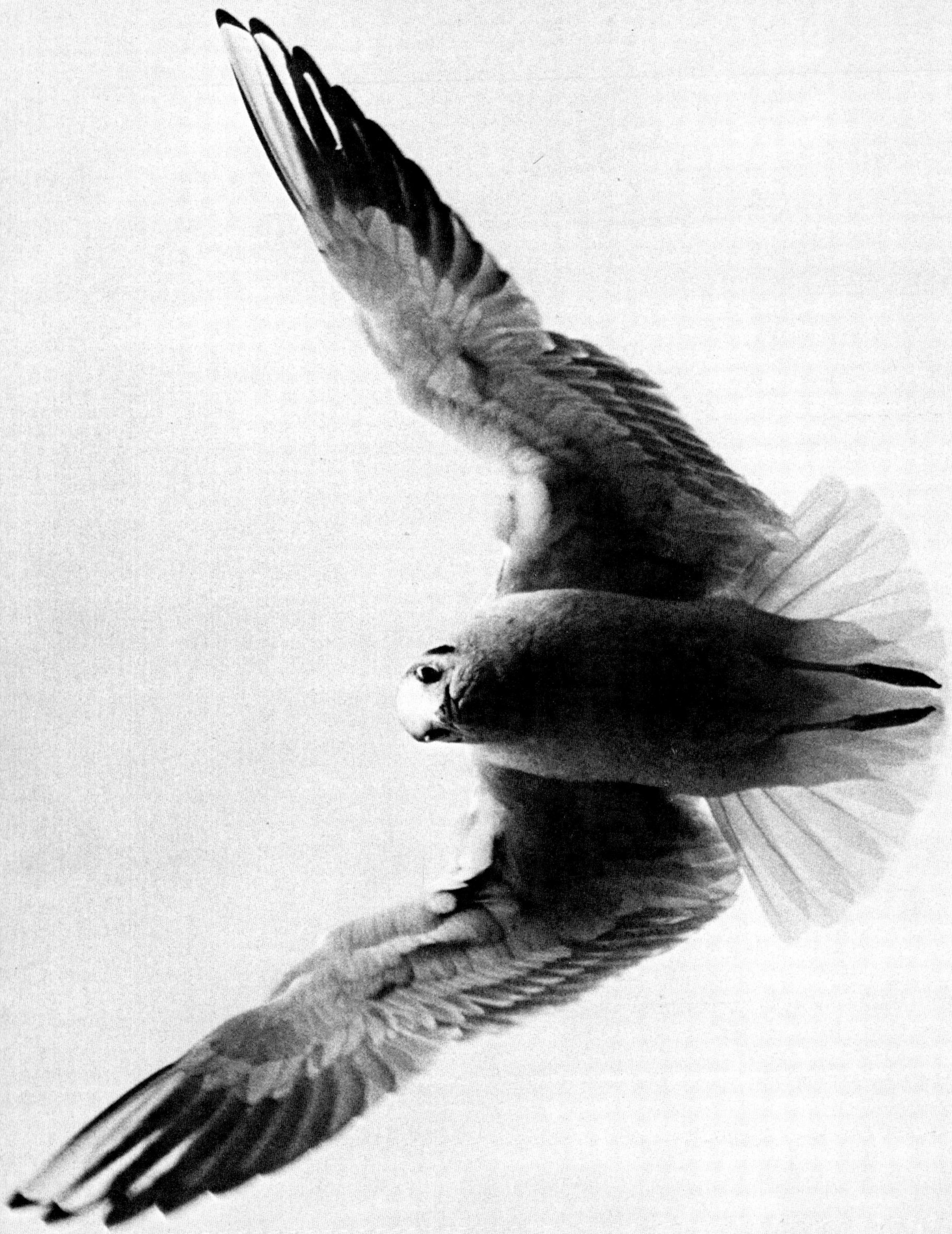

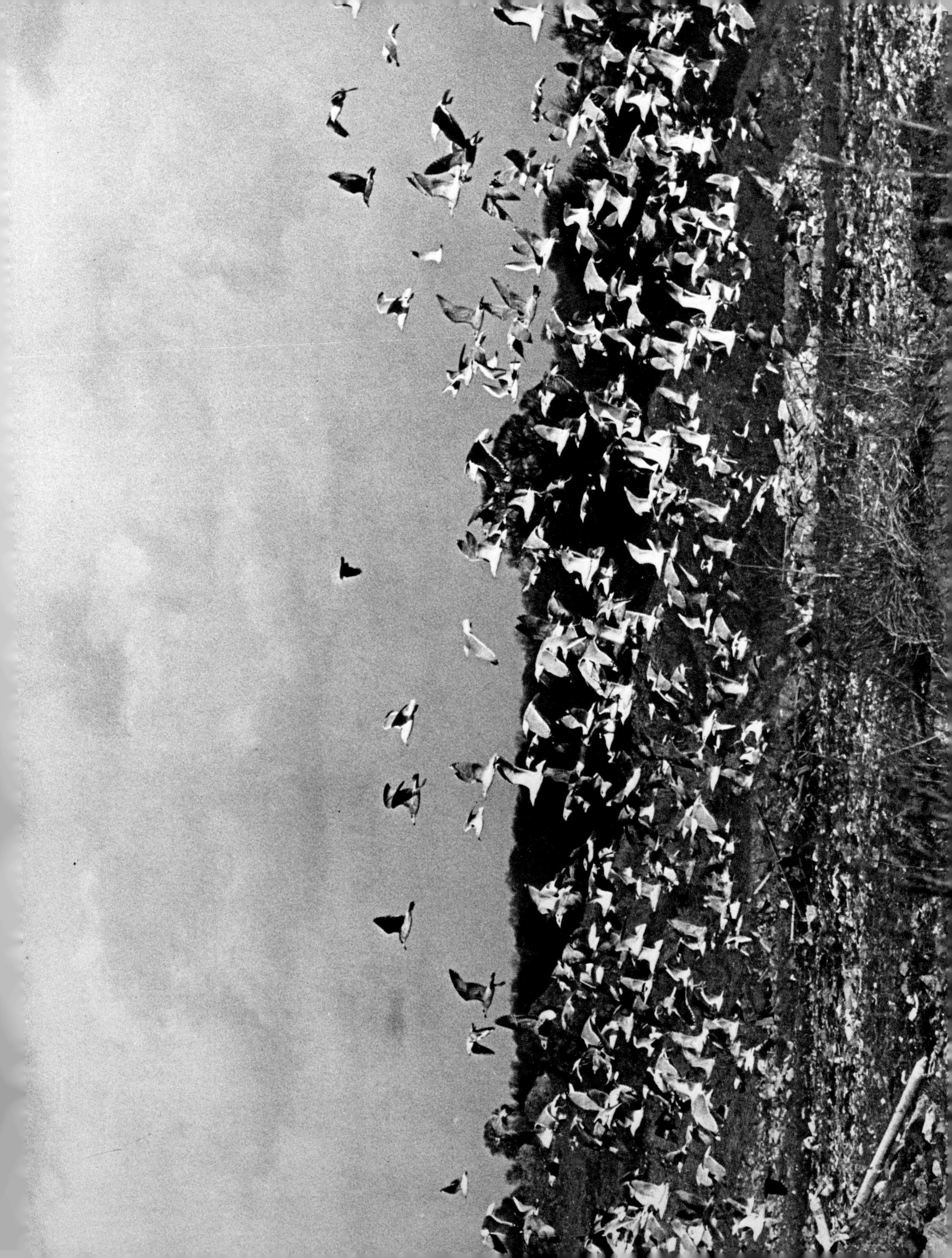

Winter Fields and Trees

The harvest has been gathered and the fallen leaves are scattered by the wind; along the margins of the fields bare trees form a bold frieze against the shifting greys of the winter sky.

   A variety of birds, their nesting completed for the year, come together to form flocks, feeding in the fields and forming characteristic patterns in the air; linear skeins of geese, erratic groups of rooks, starlings streaming overhead as they speed towards their roost, gulls forming a white wake behind the plough as they search the newly turned dark earth. These, and many more, bring life and movement to the bleak landscape of winter fields and trees.

In winter several species of grey geese migrate to the British Isles; of these, the three that occur in large numbers are the pink-footed goose, the grey lag-goose and the white-fronted goose. All these grey geese may be seen on flat shores and marshes or on fields. The pink-footed goose has already been shown mingled with barnacle geese on the coastal saltings. The photographs on this and the facing page show grey lag-geese flying from feeding in the fields to their roost on an inland loch.

When flying any distance geese are among the birds that fly in formations of vees or oblique lines. It is said that an older bird is always the leading bird, and that from time to time on a long flight the lead position will be taken by another bird. It is possible that flying in these regular formations gives an advantage to all but the lead birds; it is known that when jet fighter aircraft fly in similar formations the leading plane uses more fuel than the other members of the formation.

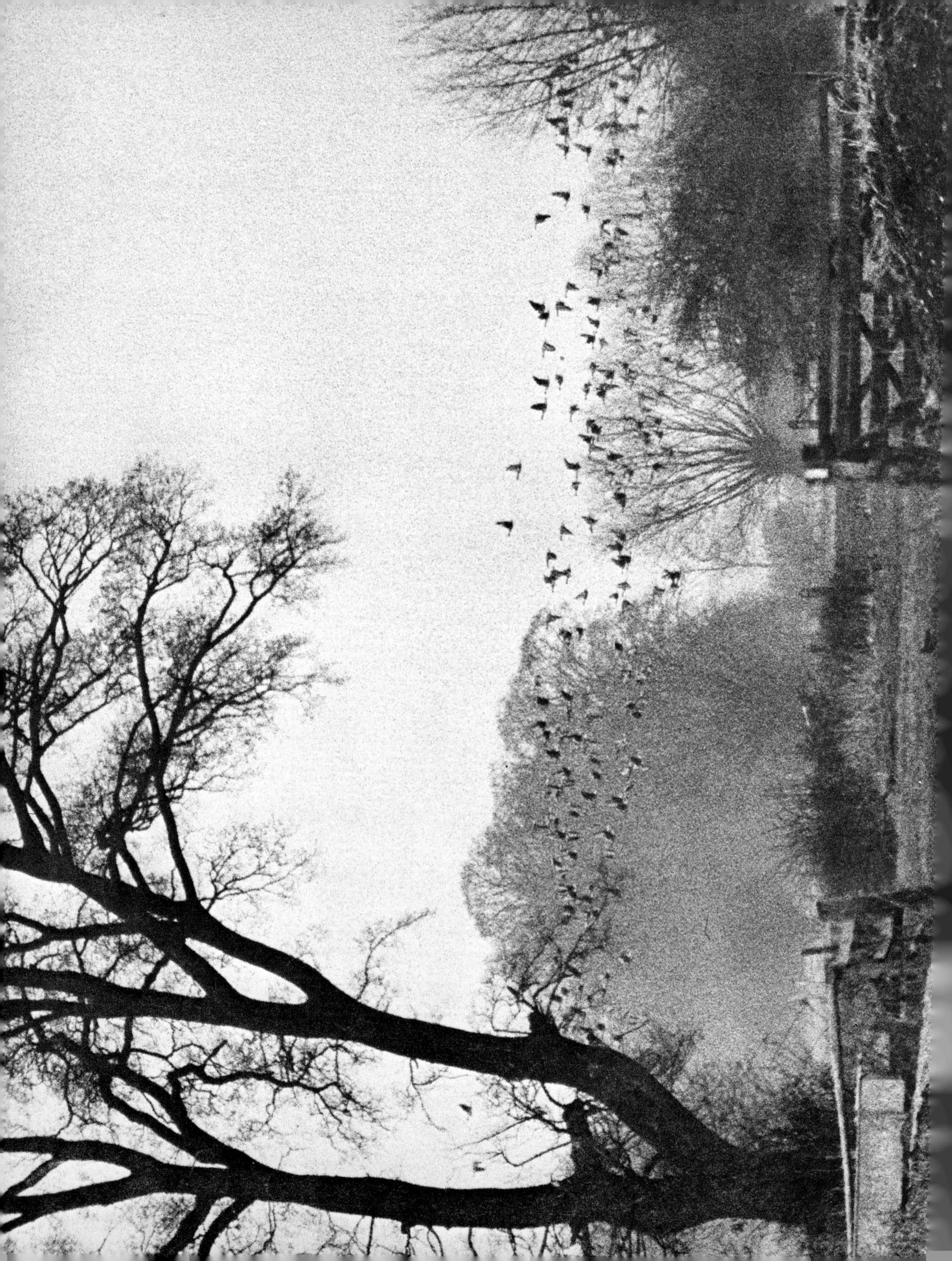

The third common species of grey goose, the white-fronted goose, is shown opposite. Part of a flock is flying from one field to another in the damp, grey light of a late afternoon in winter. The geese above, descending to land on an inland water, are Canada geese.

173

Canada geese, the great wild geese of North America, have become a familiar sight in Britain since their introduction, probably in the seventeenth century.

WINTER FIELDS AND TREES

In winter Canada geese tend to form flocks and are found on grasslands and marshes, usually near fresh water.

Flocks of starlings are seen everywhere in the fields in winter. While feeding on the ground they are fairly inconspicuous, but periodically they will all rise, exploding from the ground in a densely packed flock. Like the starlings in cities, some of these starlings will also roost socially, gathering in great numbers before nightfall.

WINTER FIELDS AND TREES

Very often starlings will gather at assembly points on their way to the roost. In this photograph starlings are arriving at such an assembly on telephone-wires. These assemblies are very noisy until the birds leave, when there is sudden silence just before they rise together to fly to the roost.

Starlings will also use telephone-wires to rest and gather upon as they roam in search of food in the winter fields.

Gulls flying from feeding in the fields pass a pylon.

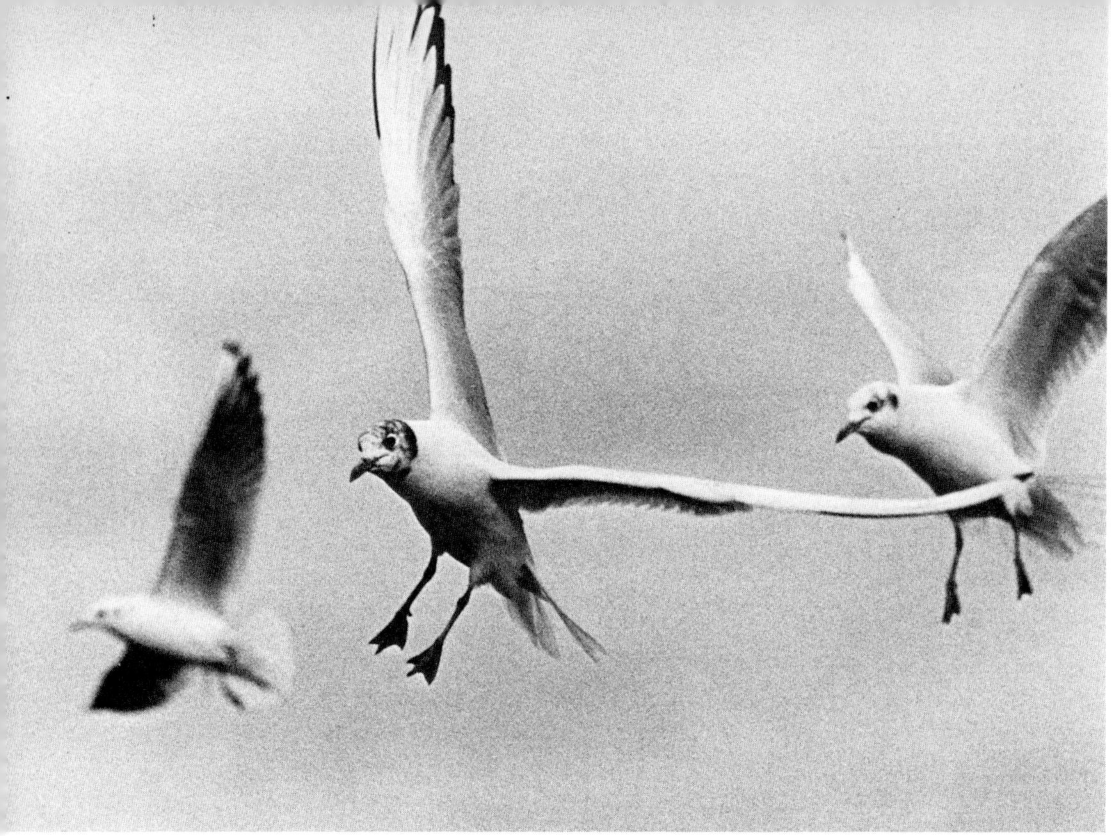

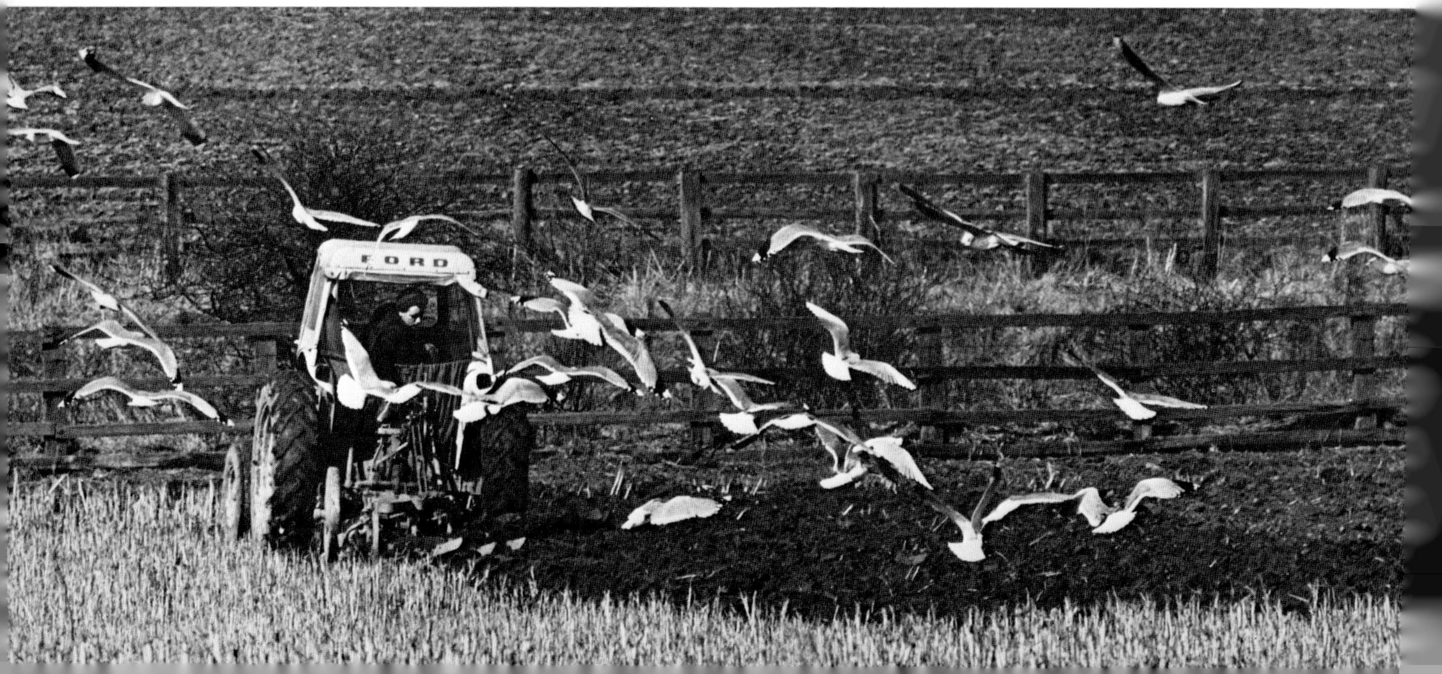
FORD

WINTER FIELDS AND TREES

Gulls are a regular sight on the fields in winter, either singly or in flocks, and the sight of gulls following a tractor is commonplace. In most of England these gulls are mainly black-headed gulls; however, the lower picture on the opposite page is of ploughing in Scotland, and as is usual there, both common gulls and black-headed gulls are searching the newly turned furrows for food.

Black-headed gulls are lighter on the wing and can hover more easily than the heavier herring gulls and lesser black-backed gulls.

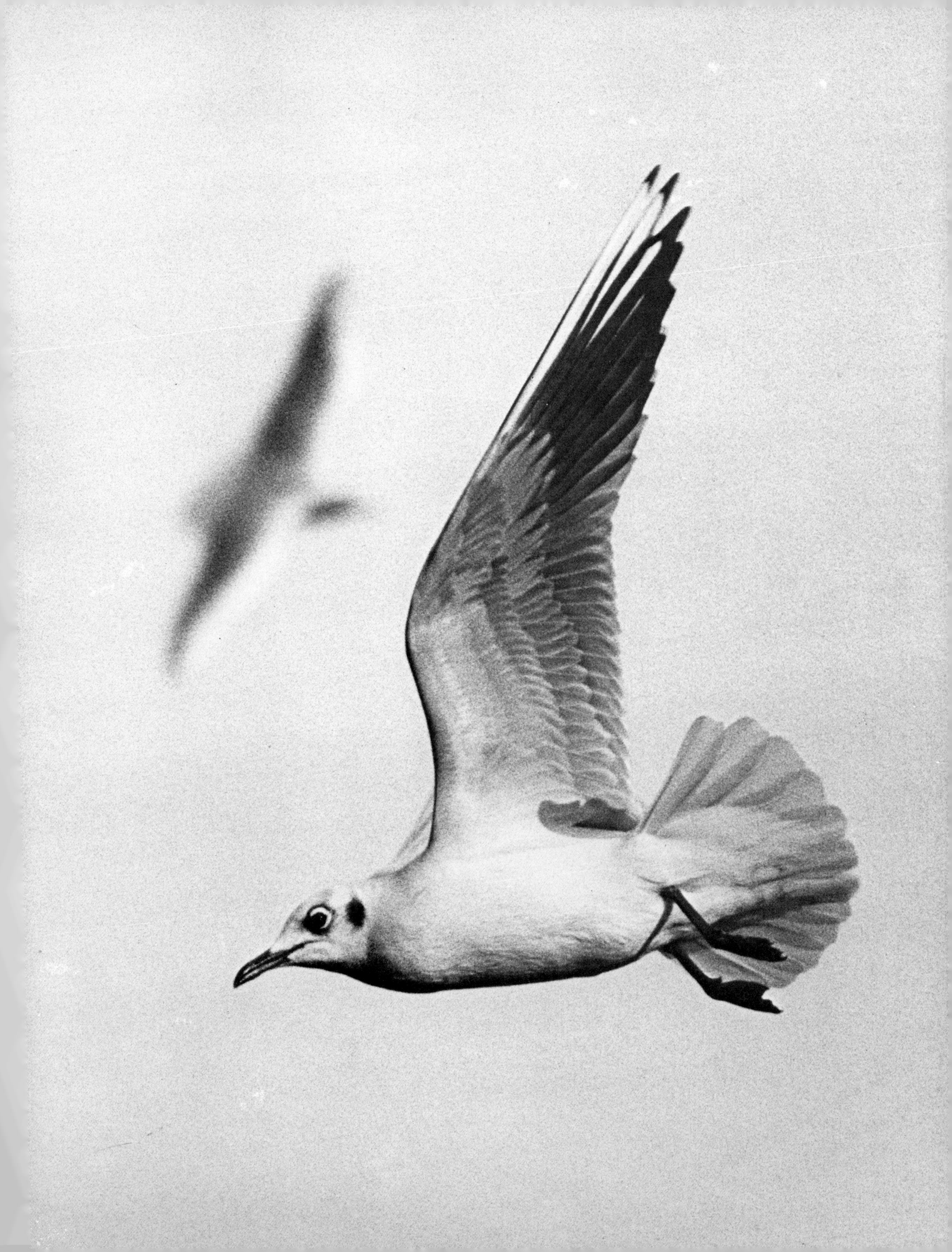

All through the winter rooks have intermingled with the other birds on shores, in the fields and on the rubbish-tips. Before winter has truly ended the rooks are flying between the bare branches to the rookery with nesting material. The winter visitors of various species are leaving, the flocks dispersing, and it is only a short while before the sea birds will return to populate the summer shores and islands.

185

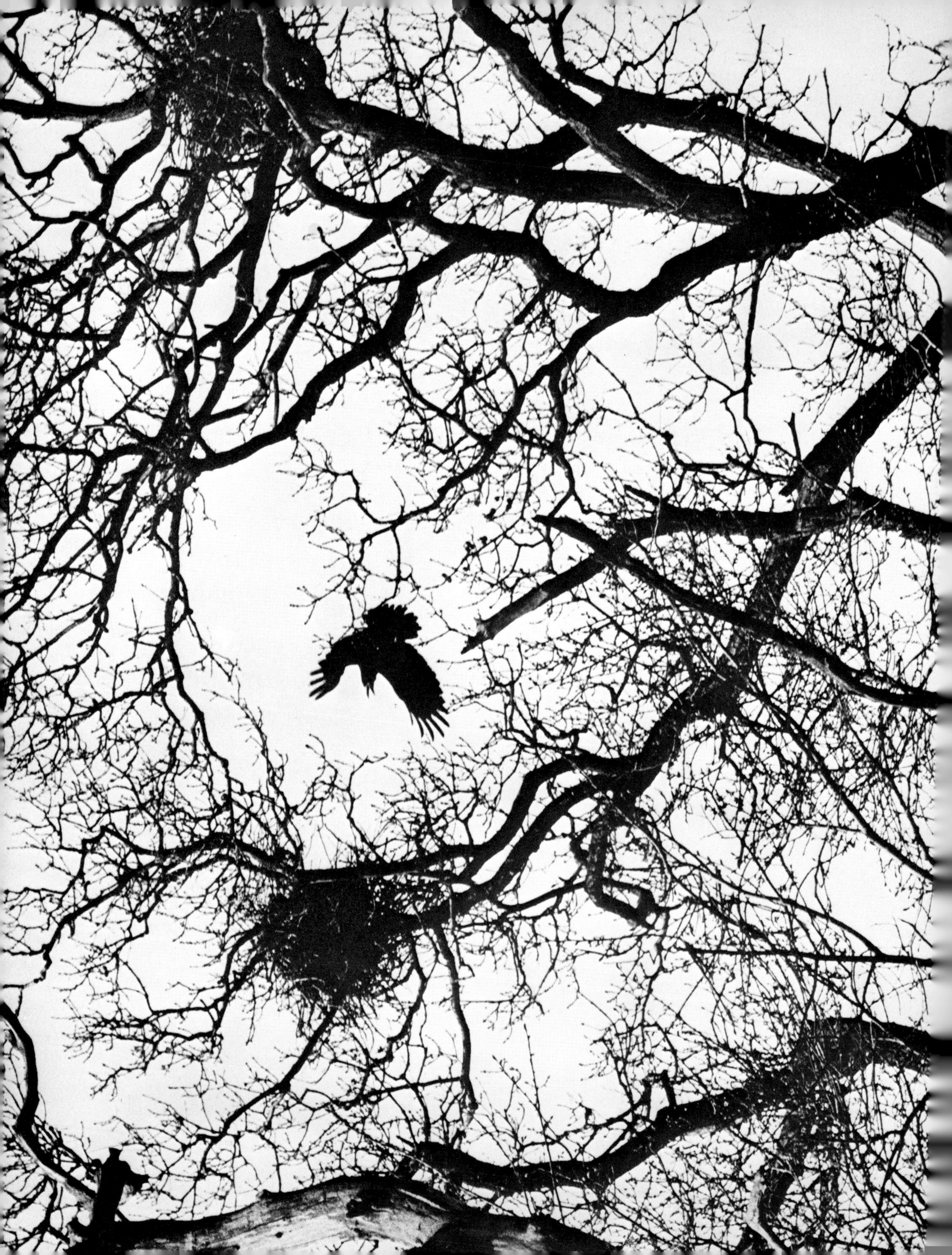

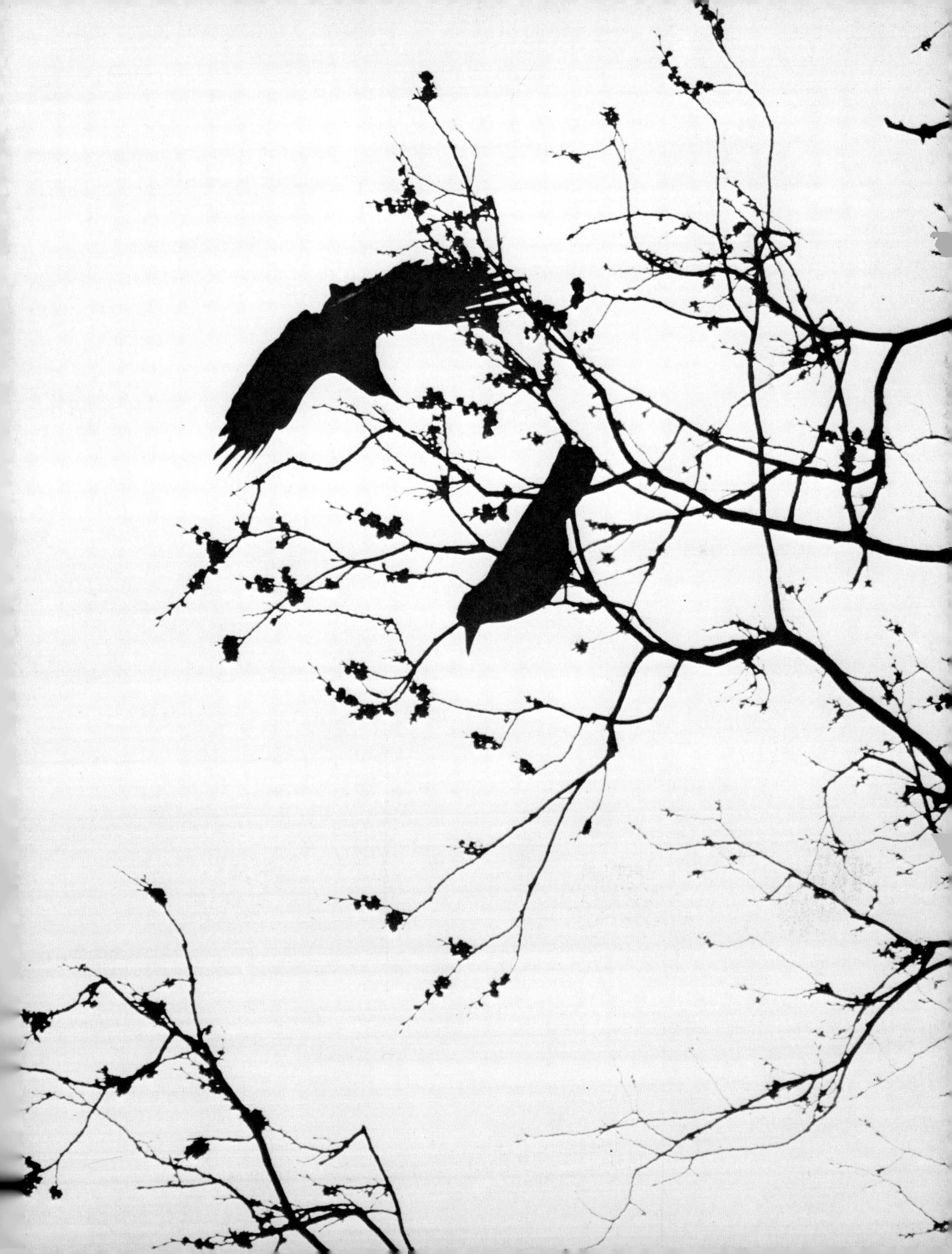

Photo Technique

The two main problems that have to be tackled in the photography of flying birds are, firstly, to obtain a large focused image of the bird in the camera, and secondly, to arrest the motion of the bird sufficiently during exposure.

The prime essential in overcoming the first of these difficulties is the use of lenses of long focal length. All the photographs of flying birds in this book have been taken with one of three lenses made by E. Leitz for the Leica camera, the 135mm Hektor, the 280mm Telyt or the 400mm Telyt.

However, although a greater advantage in terms of image size is gained with increasing focal length, the lenses also become correspondingly more difficult to use rapidly in the hand. Not only does the angle of acceptance (i.e. the view seen by the lens) decrease with increase in focal length, but so does the depth of field at any given working distance. It is necessary to practice in order to be able to find and hold a flying bird in the small field of view of one of the big lenses.

The small depth of field of the long focal length lenses demands very accurate focusing. The 280mm and 400mm Telyt lenses can be used in the Televit, a gun-like rapid-focusing mount, which greatly speeds focusing. Nevertheless, with the faster flying small birds, the most useful method is to pre-focus the camera, and wait until a bird flies into focus before exposing. As an example, a 280mm lens, set at f11 and focused at 30 feet, has a depth of field of only about 2 feet; at this distance a medium-sized bird flying towards the photographer at 20 m.p.h. will only be in focus for approximately one-thirtieth of a second.

Even with long-focus lenses it is still a necessity that the bird should fly relatively near the photographer if a reasonably large image of the bird is to be recorded on the negative. Close observation of the habits of the birds, the normal flight lines used, wind direction, and available cover for the photographer are all essential preliminaries to photography. Fortunately, many of the shyest birds, who would suffer disturbance by even a fairly close approach, fly in striking formations or tightly packed groups, and these can be photographed satisfactorily from a considerable distance. It is possible to position oneself along a flight line, so that the birds fly near, but are hardly aware of the presence of the photographer. Harmful disturbance to wild birds must always be avoided, and much damage can be done by the ignorant photographer; a photograph should never be obtained at the expense of the bird's welfare.

To obtain a sharp exposure of the bird the camera must be panned, or swung smoothly to follow the motion of the bird. Practice will determine the best stance and camera grip for each individual photographer. The weight and balance of the camera is important in this respect. The relatively light weight of a 35mm camera is one of the factors that makes it the ideal tool for this work. Even so, equipped with a long-focus lens it becomes quite a heavy piece of apparatus, and care must be taken to avoid camera shake due to jerky panning brought about by muscle fatigue. It is best to acquire the ability to hold the camera in a relaxed position at waist-level, raising it to the eye and aiming it for only a brief period before, during and after exposure.

Properly executed panning will effectively arrest the motion of the bird relative to the camera. However, the flapping wings of the bird represent motion in another direction, and this movement can only be arrested by giving a very short exposure to the film. A camera body, such as the Leicas used for the photographs in this book, with a shutter capable of exposures as short as one-thousandth of a second is necessary. Even at this high shutter speed some blur of the wings is apparent in the photographs, but a moderate degree of blur due to movement of the wings is not objectionable, and in fact helps to convey the impression of rapid movement.

In order to use such a high shutter speed with regularity, it is necessary to use film with a very sensitive, or high-speed, emulsion. Under normal lighting conditions all the year round we use Kodak Tri-X, a moderately fast film (400ASA); when working under very poor light conditions a much faster film is required, and Kodak 2475 Recording Film will then give excellent results, rated at 3,200ASA.

One of the results of using fast emulsions is the production of negatives in which the granular nature of the photographic image is obvious. We do not share the view that graininess is a bad thing in principle, but the undesired formation of excessive or poor grain can be avoided by choosing a fine-grain developer, such as Kodak Microdol-X, in association with Tri-X film. When using a film such as 2475 with a speed of 3,200ASA a high degree of graininess cannot be avoided, but a well-formed grain structure can be obtained by developing this film with continuous agitation in Kodak DK-50 Developer. A good, clean graininess can add texture and interest to the rather flat tones of a dimly lit subject.

The 35mm format is the obvious first choice for the photographer of flying birds. Not only is the camera light, but each exposure costs relatively less. With a fast-moving and unpredictable subject success cannot be achieved with every exposure; the bird may not be critically in focus, or it may have been caught in an awkward or unpleasing attitude; the dustbins of bird-flight photographers are full of such 'near misses', and the comparative inexpensiveness of 35mm film makes them less heart-rending. On the other hand, 35mm film has its various drawbacks, all relating to its small size. Since the degree of enlargement required is greater, the negative quality has to be correspondingly excellent. For the same reason, any dust or scratches on the negative will figure large on the print, and the most rigorous care must be observed in all stages of processing and negative handling.

This has been a brief description of our present methods, and it must be borne in mind that they are by no means the only way in which successful flight pictures may be obtained; photographic technique should be a progressive study, changing

PHOTO TECHNIQUE

and evolving in the light of experiment, experience and new materials. Flight photography will give great pleasure to anyone with an appreciation of natural beauty, a sharp eye and quick reflexes. The ideal picture is virtually unattainable, but the constant striving towards this goal appeals to the basic hunting instinct latent in all of us.

Index of British, American and Scientific Names

British common name	American common name	Scientific name
Puffin	Atlantic puffin	*Fratercula arctica*
Razorbill	Razor-billed auk	*Alca torda*
Guillemot	Atlantic or Common Murre	*Uria aalge*
Arctic tern	Arctic tern	*Sterna paradisea*
Gannet	Gannet	*Sula bassana*
Shag		*Phalacrocorax aristotelis*
Kittiwake	Kittiwake	*Rissa tridactyla*
Lesser black-backed gull		*Larus fuscus*
Herring gull	Herring gull	*Larus argentatus*
Oystercatcher		*Haematopus ostralegus*
Turnstone	Ruddy turnstone	*Arenaria interpres*
Knot	Knot	*Calidris canutus*
Bar-tailed godwit	Bar-tailed godwit	*Limosa lapponica*
Mute swan	Mute swan	*Cygnus olor*
Bewick's swan		*Cygnus bewickii*
Whooper swan		*Cygnus cygnus*
Barnacle goose	Barnacle goose	*Branta leucopsis*
Pink-footed goose		*Anser brachyrhynchus*
Brent goose	Brant	*Branta bernicla*
Feral pigeon	Feral pigeon	*Columba livia*
Black-headed gull	Black-headed gull	*Larus ridibundus*
Starling	Starling	*Sternus vulgaris*
Grey lag-goose	Grey lag-goose	*Anser anser*
White-fronted goose	White-fronted goose	*Anser albifrons*
Canada goose	Canada goose	*Branta canadensis*
Rook		*Corvus frugilegus*